AF659292

Propagation of Horticultural Plants

Uniform Orchard Grown from Good Nursery Stock

(*Frontispiece.*)

Propagation of Horticultural Plants

by

Guy W. Adriance

&

Fred R. Brison

BIOTECH

2018

Biotech Books

© 2018 AUTHOR

ISBN10 81-7622-052-3
ISBN13 978-81-7622-052-1

All rights reserved. Including the right to translate or to reproduce this book or parts thereof except for brief quotations in critical reviews.

The views expressed in various chapters are those of the author and not of publisher of the book.

Published by: **BIOTECH BOOKS®**
4762-63/23, Ansari Road, Darya Ganj,
New Delhi - 110 002
Phone: +91-011-23262132
E-mail: biotechbooks@yahoo.co.in

Typeset at: **Classic Computer Services**
Delhi - 110 035

Digitally Printed at: **Replika Press Pvt. Ltd.**

PRINTED IN INDIA

PREFACE

The propagation of plants has long been recognized as a fundamental practice in the fields of plant science. Various special treatments are followed in the production of horticultural plants, and for this reason their propagation presents certain problems and difficulties not encountered in other lines of work.

This book has been prepared primarily as a text for basic courses in horticulture and related courses. The essential features of plant structure and reproduction have been introduced in their relation to seed production, root formation, wound healing, and other practical phases of plant propagation. The methods of asexual propagation, including bulbs, layerage, cuttage, budding, and grafting are considered in comprehensive form. Practices followed in the propagation of certain important species are presented in detail. A discussion of certain plant diseases, and transplanting, is also included because of the relationship of these to the growth and longevity of plants (after they are planted in permanent location). Basic factors of plant growth and response are considered with respect to their relationship to commercial practices.

The material included in this work is considered to be of fundamental value to students in agriculture; it is regarded as being especially useful to students in succeeding courses in vegetable crops, fruit growing, forestry, floriculture, and ornamentals.

As a guide in practical work, the book assembles in readily accessible form recent practices that have been introduced by research workers and commercial propagators. Modifications of standard practices are reviewed and evaluated; accepted methods for the commercial propagation of specific plants are given in detail. The relative value of different root stocks, as determined by results in many parts of the country, is given considerable emphasis.

Review of the literature has been of a general nature, and no specific references have been given. Selected references to more recent publications on the various subjects have been included, with the idea that they would in turn supply references to previous publications in the same field.

The authors wish to acknowledge the assistance of Mr. S. B. Apple, Mr. R. H. Sharpe, and Mr. H. E. Wright, of the Agricultural and Mechanical College of Texas, for criticism of various chapters. Mr. Sharpe has also rendered valuable assistance in preparation of many of the photographs used for illustration. Photographs for illustrations have also been supplied from other sources, and credit for each has been given in the text.

GUY W. ADRIANCE,
FRED R. BRISON.

CONTENTS

duction of varieties on specialized rootstocks, by means of budding and grafting. These broad aspects of plant propagation are best considered from the standpoint of the technique involved in each method. A more detailed study of plant propagation should include the application of these methods to the production of specific items, such as cabbage seed, narcissus bulbs, holly cuttings, and budded roses or peach trees.

PLANT STRUCTURE

An intelligent practice of plant-propagation methods is based to a considerable extent upon a knowledge of the structure of a plant and the function of its different parts.

The larger structural units of a plant are the roots, stems, leaves, flowers, and fruits. These are known as *organs*, and each is composed of several different kinds of tissues. Tissues, in turn, are composed of cells. These are regarded as the smaller structural units of the plant. Close inspection of cells will reveal that they vary; some have thick, and others thin, walls; they differ also in size, shape, and contents.

Roots.—The root is obviously an essential organ since it absorbs water and nutrients for the plant and serves as an anchor for it.

For practical considerations the normal root of a *dicotyledonous* plant may be divided arbitrarily into two sections. Near the tip, and just behind the root cap, is a region that is characterized by certain definite qualities: (1) It is a region in which primary growth of the root takes place. (2) Branch roots arise in regular succession in the pericycle. As the region progresses outwardly as a result of elongation and growth, it is followed by secondary growth. (3) It is the region from which root hairs arise. These are relatively simple, hair-like outgrowths or extensions of the outer walls of cells of the epidermis. (4) As the root elongates near the tip, it pushes the root cap ahead of it deeper or further into the soil.

The part of a root that makes secondary growth embraces the entire length with the exception of the part near the tip. It is characterized by the following qualities: (1) The cambium ring becomes complete, whereas in the tip it exists only as strands. (2) The root increases in diameter over its entire length, but does not elongate except near the tip. (3) The epidermis and root hairs disappear on the root that is making secondary growth, and it develops a corky layer, the cortex. (4) Branch roots that form on parts of roots making secondary growth develop from adventitious buds in the cambium layer.

From the foregoing it is clear that branch roots may arise in two familiar ways—in regular succession from the pericycle

FIG. 1.—A well-developed root system showing tap and laterals. (*Courtesy, O. S. Gray Pecan Nursery, Arlington, Tex.*)

near the tip, or as adventitious roots from the cambium layer of older parts of the root. A branch root once formed in either way proceeds to develop like the parent root and soon becomes differentiated into regions making primary and secondary growth, respectively. Branch roots are produced freely on some kinds of plants; on others they are produced less readily or not at all. The roots of hyacinth apparently do not branch, and onion roots do not branch freely. Some species produce branch roots freely from the pericycle but not so readily from the cambium; others produce them freely from both tissues.

Buds and nodes are not formed on roots. When stems or shoots develop from a root, they arise from adventitious buds in the cambium. In some plants the root is an important organ

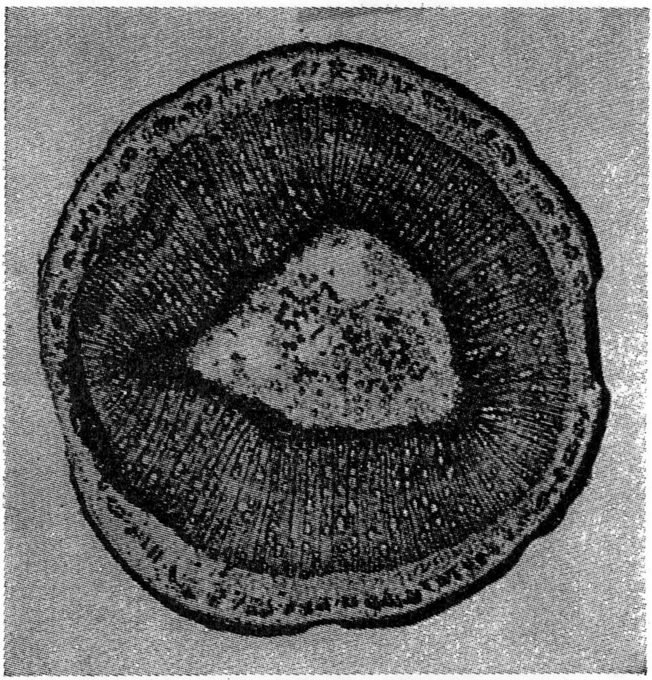

FIG. 2.—Cross section of stem of pecan, showing pith; xylem with medullary rays and water-conducting vessels; and bark, which includes cortex and phloem.

for the storage of reserve food. This function is highly developed in plants with fleshy roots.

Plants exhibit different degrees of development of the *tap* and *lateral roots*. The taproot of a young pecan tree develops more prominently than the lateral roots. In the peach and the apple, there is a very limited development of the taproot; it gives rise to laterals that comprise most of the root system. Quite naturally methods of propagation and soil and moisture considerations influence the type of root system. Rich soil normally encourages free branching, and poor soil limited branching. Plants in poor sandy soil tend to produce long roots with few branch roots.

Stems.—The stem of a plant provides the trunk and framework, which support the leaves, flowers, and fruit and expose them to the light. The trunk and framework also serve as a conducting system between roots and leaves. Stems of dicotyledonous plants have xylem, cambium, phloem, cortex, and epi-

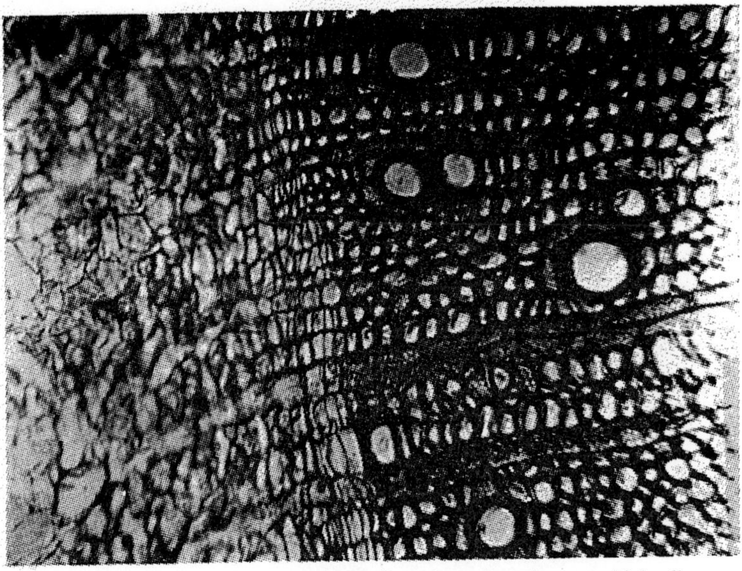

FIG. 3.—Section of pecan stem, greatly enlarged to show cambial cells.

dermis in successive order outward from the pith in the center. In old stems the epidermis and cortex have disappeared, and exposed phloem cells form the bark.

Stems of dicotyledons make terminal growth by elongation of cells near the tip of the growing branch. A plant increases in length or height as long as it lives. Increase in diameter takes place largely after the stem begins secondary growth. Meristematic cells located between the xylem and the phloem of a stem are variously designated as the *cambium*, the *cambium layer*, and the *cambial region*, depending upon whether it is thought to be a meristem without regard to thickness, a single layer, or several layers of cells in radial thickness. Regardless of which is true, the cambium serves the tree in several very important ways: (1) It is the meristematic tissue responsible for increase in diameter or circumference of stems. Cambium cells during each season of growth enlarge and divide, and give rise to new cells toward the inside and toward the outside. Those toward the inside

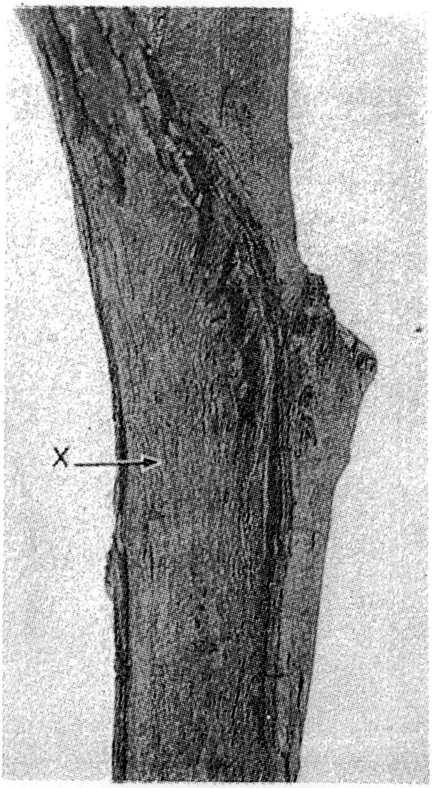

Fig. 4.—The tissue shown in the area marked by the *X* developed by regeneration from the injured cambium when the original bark was peeled off.

differentiate into xylem tissue cells of one kind or another; those formed on the outside produce the phloem. (2) It is the tissue that enables wounds to heal. This is accomplished by two familiar processes. *Regeneration* may take place where bark is removed and living cambium cells are exposed on the surface of the wood. If weather conditions are suitable, such cells may become active and reconstruct a bark. *Overwalling* takes place as a result of the activity of cambium cells around the margins of a wound, causing new tissue to advance from various sides and cover the wound. (3) The cambium produces the outgrowth of young, tender, callus tissue, which is so essential to the success of budding and grafting. The same tissue forms

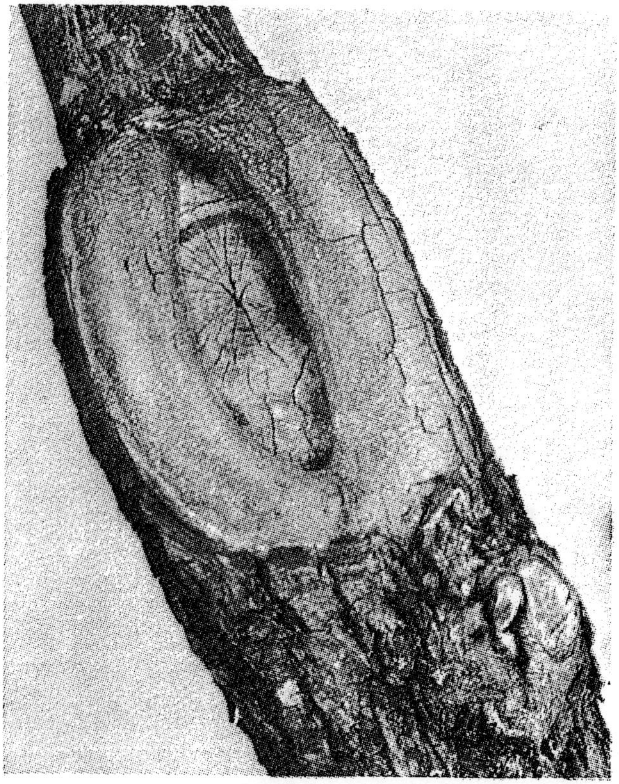

FIG. 5.—Showing the healing of a wound by overwalling. No wound compound was applied and the wood checked as a result of exposure.

on the cut ends of cuttings of some plants and apparently seals them against destructive organisms. (4) When roots develop on stems, as in cuttings, they arise in most instances directly from adventitious buds in the cambium region, although in some cases they develop from preformed initials known as "root primordia," present in the cambium region.

Stems of monocotyledonous plants have the cambium strands and vascular bundles isolated in masses of other tissues.

Buds.—Young stems are characterized by the formation of *buds* at points that are called *nodes.* The area between nodes is known as the *internode.* A bud is a growing point surrounded by embryonic leaves or flower parts. It is in reality a rudimentary stem, protected by an envelope of scales. Close examination of a bud reveals leaves and buds in the same order as on a growing

FIG. 6.—Buds on 1-year stems. 1, Hickory, showing (1*a*) terminal bud and (1*b*) lateral bud; 2, Pecan, showing (2*a*) lateral bud in terminal position; 3, Pecan, showing (3*a*) primary bud and (3*b*) secondary bud.

stem of the same plant. Several classifications of buds are recognized. They are based principally on the mode or time of origin, position on stem, position at node, time at which they begin growth, and function.

1. *Position on Stem.*—As a result of the growth of a bud a new shoot forms and the bud in question ceases to exist as such. In

its growth, however, the new shoot forms *lateral* buds at various nodes and when growth stops at the end of the season a *terminal* bud may or may not form, depending upon the species or variety. In the event a terminal bud does form it is usually the one to begin growth first the following spring. If a terminal bud is not

FIG. 7.—Showing blind buds of peach, between terminal and lateral buds which had begun growth.

formed, future growth of the twig is from lateral buds, usually near the apical portion of the shoot.

Most lateral buds are borne in the axils of leaves, and hence are known also as *axillary* buds. In some species of sumacs the leaf petiole surrounds and encloses the bud. Buds may occasionally occur at points along a stem other than in leaf axils, in which case they are classified simply as lateral buds. Examples of the latter class are frequently to be observed on parts of pecan shoots that are formed near the end of the growing season.

The peach, tung tree, and less frequently the rose produce shoots with certain nodes and leaf axils at which no buds occur. These are known as *blind buds.* It would be more appropriate to designate them as *blind nodes.*

2. *Position at Node.*—In some species only a single bud develops in the axil of each leaf. In others, two, three, or even five may develop. From two to four buds are commonly formed at a node of the pecan or walnut. The one nearest the terminal of the shoot is usually the largest of the group and is referred to as the *primary* bud. The next bud then becomes the *secondary* bud, the third one the *tertiary* bud, and so on. Commonly, however, all the group except the primary bud is referred to collectively as the *secondary* or *reserve* buds.

The primary bud is the one most likely to grow when the tree initiates growth in the spring. The reserve buds also may be forced into growth in the event of excessive rainfall, frost, or insect damage to the growth from primary buds.

3. *Function.*—Fruitgrowers often speak of *leaf buds* and *fruit buds.* Leaf buds contain a growing point of undifferentiated tissue. They give rise only to new vegetative shoots. The upward and outward extension of a tree is due largely to the growth of leaf or vegetative buds.

Fruit buds contain the rudimentary blossoms with various parts of the flower enclosed. Under favorable conditions they give rise ultimately to fruit. These buds appear on plants in the same general position as leaf buds. This would be expected, since they differentiate from a previously developed leaf bud. The change from leaf bud to fruit or flower bud takes place in some species during the season previous to the one in which the flowers appear. In the peach, for example, the fruit buds that are to appear in the spring are formed during the previous summer and fall. In other species the factors that determine the formation of fruit buds are apparently operative for only a short period prior to the time at which the buds begin to grow. The pistillate flowers of the pecan, for example, do not differentiate until the approach of the growing season in the springtime. Fruit buds of citrus are also differentiated in late winter or early spring just preceding their appearance.

Some species of plants produce mixed buds, containing both flowers and a vegetative axis within the same cluster of bud scales. Fruits of the apple and pear are produced from mixed buds. The catkins and lateral leaf buds of the pecan are borne together, enclosed in the same scales; when growth begins in the spring the leaf bud starts growth and either develops into a new shoot, or falls off after having made limited growth. In either case the catkins are exposed, develop, shed their pollen, and ultimately fall off.

4. *Dormant and Latent Buds.*—The buds of most fruits develop and mature during a given season and remain dormant over winter. With the initiation of growth the following spring such buds either (*a*) grow into twigs, (*b*) begin growth and later fall off, or (*c*) remain dormant. Those that remain inactive long past the time at which they might have grown become "latent" buds. They may even become buried in layers of bark, but later, perhaps several years, push into growth, generally as "water sprouts." Usually, however, these latent buds make sufficient annual growth outward to prevent them from being overwalled by bark or other tissue.

5. *Adventitious Buds.*—Occasionally, young shoots or twigs appear at points other than nodes. They are neither terminal on stems nor lateral in the axils of leaves. These are called *adventitious shoots* and are considered to have developed from adventitious buds. Such buds originate by the differentiation of cambium cells into a growing point. They may develop on a root or a stem. Unusual conditions as, for example, excessive soil moisture, defoliation by insects or freeze, or mechanical injury caused by cultivation or by pruning, may stimulate the development of adventitious buds.

It was stated in a previous paragraph that root growth from a stem is in some cases from adventitious buds, although in other cases roots arise from so-called "root primordia." Shoots arise from roots invariably from adventitious buds, since no other kind of bud is present on roots. Secondary growth of roots is characterized by completion of the cambium ring. Branch roots arise from portions of roots making secondary growth only from adventitious buds. They are referred to as *adventitious roots,* in order to distinguish them from roots that arise in regular succession in the pericycle of primary roots.

It is of incidental interest that the pith of shoots or roots that develop from adventitious buds, from either roots or stems, extends only to the point of origin of the bud. The pith of shoots that develop from true buds has an uninterrupted connection with the pith of the mother branch, perhaps maintained across a dozen or more annual rings. And the pith of roots that develop in regular succession extends to the pith of the mother root.

Species of plants, and even varieties, differ in the readiness with which growth is made from adventitious buds, either on the stem or roots.

Suckers and Water Sprouts.—Frequently shoots that make a very rapid and luxuriant growth develop around the bases of plants. Such sprouts that grow from adventitious buds on the roots are known as *suckers*. They form freely in some species. In the blackberry they may be used in multiplying plants. They occur freely in the native persimmon, sumac, elderberry, and some plums. The readiness with which suckers form has an important bearing on propagation practices for a species. The Cherokee rose (*Rosa laevigata*), for example, has many characteristics that commend it as a budding stock for standard varieties; yet it is never used as such because of its persistent habit of suckering. Rupestris St. George grape is a valuable stock for grapes, but in some sections it is objectionable because of its free-suckering habits. The Mustang grape (*Vitis candicans*) is subject to the same criticism. Plants that produce suckers freely may usually be propagated readily by root cuttings.

Vigorous-growing shoots that arise from the trunk or large limbs of trees are called *water sprouts*. They may develop from adventitious buds but usually come from true buds that have remained dormant long past the time when they might have begun growth. Water sprouts occur also on herbaceous plants such as the chrysanthemum.

Suckers and water sprouts are commonly regarded as being unfruitful, but with age they may become as fruitful as other parts of the tree.

Leaves.—Mineral nutrients and water from the soil are combined with carbon dioxide in the leaves, under the influence of sunlight, to form plant foods essential to growth. Leaves are lateral appendages formed by the stem in elongation. Plants that shed their leaves at certain seasons, and hence have a period during each yearly cycle when they are bare and another when they are in full foliage, are known as *deciduous* plants. Examples are peach and pecan. Those that retain their leaves for long periods, and do not shed all of them at one time but shed them so gradually that the trees have leaves on them at every season of the year, are known as *evergreen* plants. *Broad-leaved* evergreens are represented by such plants as citrus and cherry laurel, which retain their leaves for more than 1 year, and by others such as the live oak and yaupon, which retain them only until new leaves are formed the following spring. The pines, arborvitaes, and junipers are examples of the *coniferous* evergreens.

Flowers, Seed, and Fruit.—The flowers, seed, and fruit are highly specialized parts of the plant structure. They will be treated separately in chapters that follow.

CLASSES OF PLANTS

Life Cycle of Plants.—The life cycle of any plant is the entire process of growth that is involved from the germination of the seed to the production of a crop of seed by the new plants. The actual duration of this period is very variable, since some plants mature and produce seed much more quickly than others. Portulaca and some of the grasses finish their life cycle in a few weeks, while the Northern Spy apple does not bear fruit or make seed until 10 to 15 years old.

The behavior of certain plants, from the standpoint of the life cycle, is quite different from that of others. One group of plants, known as *monocarpic*, produces only one crop of seed and then dies. In many of these plants the length of life may be prolonged by preventing them from flowering and producing seed. Other plants, which may also be said to have completed their life cycle when they produce seed, do not die but continue to live and produce seeds for many years.

Annuals.—These are plants that mature seed during the same season in which the seedlings are started. Many annuals, such as zinnia, produce their seed early enough to make a second stand of plants in the latter part of the growing season. Corn, cotton, and watermelon are examples of the annual plants.

Some plants fall into this group of annuals as the result of certain conditions of environment. Cotton is a plant that lives many years in its native environment but is an annual under the climatic conditions of the United States and seldom lives longer than one season. The papaya in Texas and Florida also becomes an annual in some cases.

Plants from other groups may come in the annual group under other conditions. With the long growing season of the South,

FIG. 8.—Pecan tree in Tennessee with trunk diameter of 7 feet and limb spread of 120 feet. (*Courtesy, Department of Conservation, State of Tennessee.*)

many biennials complete their life cycle in one season. Other biennials, of which Canterbury Bell is a notable example, have been changed over to annuals by the efforts of plant breeders.

Biennials.—Plants that make one season's growth, survive the winter in the dormant condition, and produce seed stalks and seed the second year are known as *biennials*. The entire plant may go through the winter, or it may die back to the roots, and come up again. The plants normally produce seed and die before the end of the second season of growth. In many cases, biennial plants run to seed the first year, and do not last over into the second season. Mustard, cabbage, and hollyhock are examples of biennials.

Perennials.—These plants persist from year to year, often not producing seed for many years, but not usually dying immediately after seed production. Fruit orchards, especially nut orchards, produce seeds year after year, over a long period of time. There are three general classes of this group:

1. *Herbaceous perennials* have tops that die down every season, while the roots persist. These plants are typified by mallow, asparagus, and Queen's crown vine.

2. *Woody perennials* have a perennial root also, but the tops last two seasons instead of one. Dewberries and blackberries are of this type; the canes that are produced one season will fruit and die the next.

3. *Trees* and *shrubs* are completely woody plants, which persist from year to year, adding to that portion of the plant body previously produced. Each of the annual rings in the wood of trees cut transversely normally represents 1 year's contribution to the body of the plant.

Questions

1. Name the three principal phases of horticulture.

2. Define plant propagation.

3. Why is the fruitgrower concerned with plant propagation?

4. What are the larger structural units of a plant?

5. What are the chief functions of the roots of a plant?

6. Name functions of the part of a root near the tip that makes primary growth.

7. Name functions of the part of the root that is making secondary growth.

8. How do root hairs arise? What is their function? How long do they live?

9. In what ways do branch roots arise?

10. What is the function of the stem of a plant?

11. Where is the cambium located? Of what value to the plant is it?

12. What is callus? Regeneration? Overwalling?

13. Define a bud.

14. Classify buds according to position on stem, position at node, function, etc.

15. What is the difference between a sucker and a water sprout?

16. Distinguish between deciduous and evergreen plants. What are common examples of each kind?

17. What is the meaning of the term *life cycle?* What are the classifications of plants from this point of view? Give examples of each class.

Suggested References

1. AUCHTER, E. C., and H. B. KNAPP: "Orchard and Small Fruit Culture," pp. 185–196, John Wiley & Sons, Inc., 1937.

2. GARDNER, V. R., F. C. BRADFORD, and H. D. HOOKER: "Orcharding," pp. 1–21, McGraw-Hill Book Company, Inc., 1927.

CHAPTER II

SEEDS

Seeds are extremely variable in size, shape, color, and longevity. Certain flowers and vegetables produce seeds that are as small as grains of sand, and some are dust-like. The avocado seed may be as much as 2 inches in diameter, and the seed of the coconut is much larger. In shape and texture seeds may be thin, flat, rounded, wrinkled, angled, smooth, or irregular. With all these variations, true seeds have three essential parts in common, which become their distinguishing characteristics: (1) The *embryo* is the most important constituent of a seed. It is a living plant whose growth has been restricted by the maturity of the seed. (2) Another important characteristic of a seed is the *stored food* accumulated in the seed before it was separated from the mother plant. This reserve food may be contained within the embryo or in the endosperm, which closely envelops the embryo of certain classes of seeds. (3) The *testa* is the outside covering and protective coat of a seed. It is formed from the integuments of the ovule.

Structure of Flower.—The flower is the forerunner of a seed. In order to consider the processes that result in seed formation, it is important to give some thought to the structure of flowers. There are two essential parts of the flower: the *pistil* and the *stamen.* The pistil consists of the stigma, the style, and the ovary, which is the lower, enlarged portion. In it are borne the ovules, which, when mature, become seeds. The style forms the connection between the ovary and stigma, and through it the germ tube passes on its way into the ovary. The stigma represents the upper portion of the pistil; it receives the pollen and affords a favorable medium for its germination. The stamen is made up of the filament, or stalk, and the anther, in which the pollen grains are produced. Enclosing these essential parts in two outer whorls are the *corolla* and the *calyx.* The corolla is made up of an indeterminate number of petals. The calyx is composed of sepals. The calyx and corolla are accessory parts but not essential to the formation of seeds. It is considered that their function is to attract insects, some of which are helpful in pollination; they possibly afford protection to some flowers. The calyx and corolla are not present in some flowers; in others they are fused to form a *perianth*, as in the pecan. Two general classes of flowers are recognized:

1. *Perfect.*—Those flowers that have both stamens and pistils and are, hence, generally capable of self-pollination are known as *perfect* or *hermaphroditic* flowers. Perfect flowers are borne in most of the common fruit and vegetable plants. Such fruits as

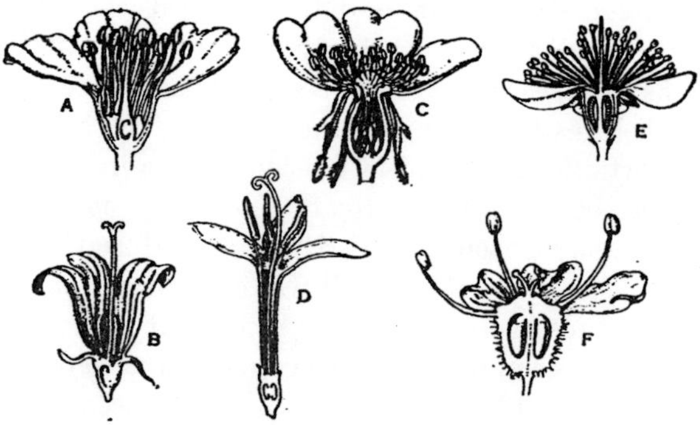

FIG. 9.—Vertical sections of several representative types of flowers. *A*, almond, *Amygdalus* (*after Le M. and Dec.*); *B*, bluebell, *Campanula* (*after Baillon*); *C*, rose, *Rosa* (*after Le M. and Dec.*); *D*, coffee, *Coffea* (*after Baillon*); *E*, myrtle, *Myrtus* (*after Le M. and Dec.*); *F*, carrot, *Daucus* (*after Baillon*).

the apple, peach, plum, orange, grapefruit, and lemon have perfect flowers; the tomato, bean, lettuce, garden pea, onion, and many other vegetables likewise have perfect flowers.

2. *Imperfect.*—Species that have only stamens or pistils within a flower are divided into two groups, on the basis of the distribution of the flowers on plants in each group:

a. Monoecious species are those that have the stamens and pistils borne on the same plants but in different flowers. The pecan, hickory, and walnut, which bear long, pendulous catkins of staminate flowers, usually on 1 year-old wood, and terminal clusters of pistillate flowers on new growth, are good examples of this case; the filbert, tung tree, and the oaks are other examples. The cucumber, squash, pumpkin, watermelon, and sweet corn are species of vegetable plants that are monoecious.

b. Dioecious species are those that bear stamens and pistils on separate plants. The date palm, Smyrna fig, Capri fig, asparagus, spinach, and common American persimmon are examples. The commercial varieties of Muscadine grapes are likewise dioecious in effect. Certain of these plants produce staminate flowers that are apparently hermaphroditic flowers with the pistils

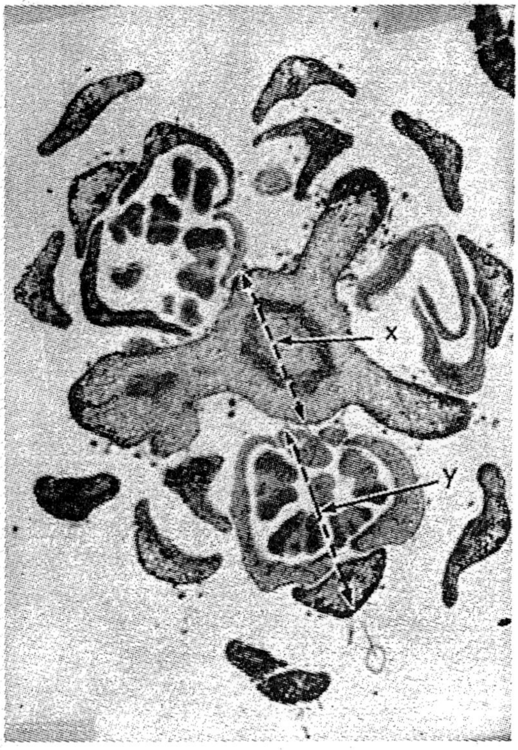

FIG. 10.—Cross section of a pecan catkin, showing peduncle at *x* and anther at *y*.

almost or wholly suppressed; and the pistillate vines produce flowers with pistils that are normal, but the stamens are more or less recurved and rudimentary. The strawberry shows some tendency toward the dioecious habit; some varieties of it bear perfect flowers and others bear pistils and only rudimentary stamens. Since staminate plants of the strawberry are not fruitful, they are usually discarded, and perfect-flowered varieties are used to pollinate the pistillate forms.

Pollination.—The first essential step in the process of fertilization is known as *pollination.* It is merely the application of pollen to the stigmatic surface. When the anther is mature, it splits along longitudinal grooves and the pollen grains are discharged. The process by which pollen is shed is known as *anthesis.*

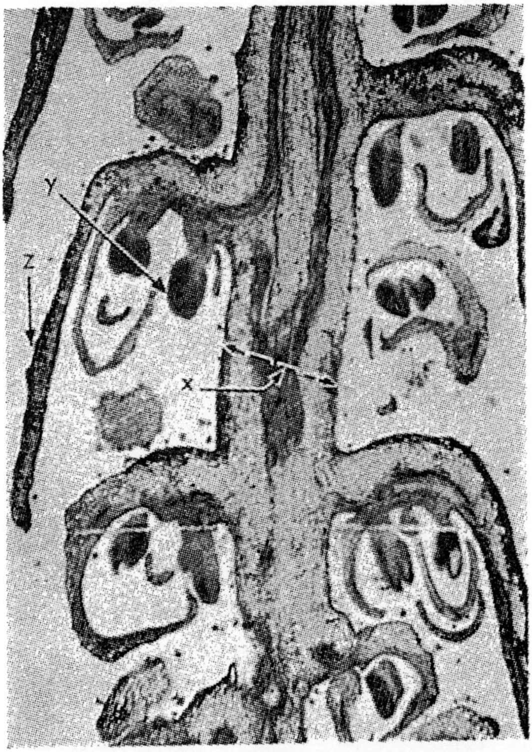

Fig. 11.—Longitudinal section of a pecan catkin with peduncle indicated at x; bract at z, and anther at y.

1. *Self-pollination,* or *autogamy,* is the pollination of a flower with its own pollen. The structure of flowers that are normally self-pollinated determines the extent to which pollen may be brought in from other sources. In the bean, for example, in which the pistil and stamen are enclosed in a complex floral arrangement, there is little opportunity for cross-pollination. In other cases, such as the apple and the tomato, flowers may be self-pollinated or easily cross-pollinated with pollen from other

sources introduced by bees. With such flowers, it is a question of chance whether the pollen that furnishes the sperm cell for fertilization comes from the same or a different flower.

2. *Cross-pollination*, or *allogamy*, is the pollination of a flower with pollen from any other flower, whether on the same or another plant. Cross-pollination between two flowers of the same plant is known as *geitonogamy;* there is no difference from the genetic viewpoint between seed resulting from this process and those resulting from true autogamy. Neither is there any difference from the genetic viewpoint between seed resulting from cross-pollination between flowers of two asexually propagated plants of a variety and those resulting from self-pollination. Plant breeders generally regard self-pollination as the transfer of pollen from the anther to the stigma of the same flower or from one flower to another of the same plant, and cross-pollination as the transfer from one plant to another.

Agents of Pollination.—There are various ways by which pollen may be transferred from the anther to the stigmatic surface of the pistil.

1. *Growth Processes.*—In some plants the pollen is brought into contact with the stigmatic surface of the pistil in the process of the growth and development of the flower. For example, in lettuce, at the beginning of anthesis, the pistil grows through the anther tube and pollination takes place.

2. *Insects.*—Flowers of some plants have large showy petals and nectar glands within the flower that attract insects. The pollen grains of such flowers are generally sticky, heavy, and, in some cases, covered with an oily film. This is true of the pollen of flowers of such common plants as the peach, plum, cherry, apricot, apple, pear, tomato, and muskmelon. As a result, the grains adhere freely to one another and to the bodies of insects that visit the blossoms.

Insects are important in the pollination of both self- and cross-pollinated flowers. Self-pollination of hermaphroditic flowers of the tomato, for example, is aided if the flowers are visited by insects. Because of the stickiness of the pollen, insects are helpful in transferring the pollen even the short distance from the anther to the pistil.

The fig wasp is important in the pollination of certain kinds of figs. In recent experimental work in onion breeding, flies have been used as agents of pollination of plants that are grown in wire cages. Many different kinds of butterflies and bees are active in visiting flowers, and in doing so they act as agents in the distribution of pollen. Of all the insects, common honey bees are far the most important for pollination purposes. They are best adapted by structure of the body to act as carriers of pollen; they survive the winters in great numbers and are active in early spring when fruits bloom; they are constant for only one kind of a flower at a time; and, finally, they can be moved about and made available wherever they are needed.

Fruitgrowers in sections where there are few wild bees place hives of bees in the orchard during the period in early spring when they are needed for pollination purposes. In the greenhouse culture of American varieties of cucumbers, special provisions must be made for pollination. The pistillate flowers may be hand-pollinated, but a more successful way involves the use of bees. The hives are kept either on the outside of the greenhouse, usually with a pane of glass removed, and individual hives so placed that the bees enter the house through the opening; or the hives may be located inside the house. A very large greenhouse might require several colonies.

3. *Wind.*—Some species of plants are pollinated by wind. The date palm, pecan, walnut, filbert, corn, spinach, and beet are examples of plants of this class. These plants do not have the conspicuous flower parts that characterize those pollinated by insects. The pollen produced by them is very fine, light, dry, and easily borne by the wind. Investigations have shown that pecan pollen may be blown 3,000 feet, and it is generally believed that it may be blown much farther. Rain, dew, and fog inhibit the distribution of pollen by wind. It has been found that no shedding of pollen occurs in the pecan when the relative humidity of the air is above 85 per cent. Prolonged periods of rainy or extremely humid weather thus tend to reduce fruitfulness in wind-pollinated plants. Such unfavorable weather during the pollination period may account, in some degree, for the uncertainty of crops of wind-pollinated species in sections where such conditions are likely to occur. With some plants, both wind and insects are responsible for pollination. Such is the case with grapes.

4. *Other Agencies.*—Water, birds, snails, and thrips are also agents of pollination in special cases.

Fertilization.—The stigmatic surface of the pistillate flower, when the latter is receptive, usually has a sticky or viscid fluid upon it. The pollen grains, falling on this favorable medium, germinate in a comparatively short time, and a tube grows out from one of the pores of the grain. This tube, known as the *pollen tube*, penetrates the style of the flower, supposedly by dissolving its way between the cells, and finally penetrates the ovule. As it grows, the nucleus of the pollen grain, which has divided into two parts, follows down the tube. When the tube finally emerges into the ovule, the two nuclei are discharged into the embryo sac. The time required for this process varies, commonly, from a few hours to as much as 5 days; in the pecan it requires from 2 to 3 weeks, and in the overcup oak it requires 1 year.

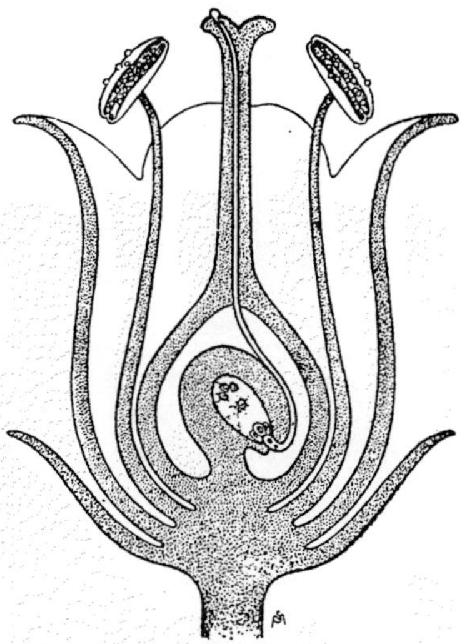

Fig. 12.—Diagram to show the general anatomy of a complete flower and the process of fertilization.

In the meantime, within the ovule, certain changes have taken place that finally result in eight nuclei, arranged within the embryo sac. These consist of the egg cell and two synergids at one end, three antipodal nuclei at the other end, and two polar nuclei near the center. As the two generative nuclei are discharged from the pollen tube, one of them fuses with the egg cell, producing the embryo of the seed. The other generative nucleus usually fuses with the two polar nuclei, to produce the endosperm. This complete process is known as *double fertilization*.

In the first growth of the embryo after fertilization it becomes differentiated into certain regions. It is held in place by a suspensor, which may consist of only three or four cells or may be only one cell wide and very long. Next in order of development are the radicle, which is the root in the young plant; the hypocotyl, which is the lower stem; the epicotyl, which is the stem between the cotyledons and the first true leaves; the cotyledons, or seed leaves; and the plumule, or growing point, of the young plant. As the seed develops and matures, food materials in concentrated form are deposited within the seed coat. These consist of carbohydrates, proteins, and fats in varying proportions, and different kinds predominate in different seeds. In some seeds stored food is largely confined to cells of the embryo; in others it is in the endosperm adjacent to the embryo.

Classes of Horticultural Plants.—Interest in pollination and fertilization is quite naturally determined by the phase of the

FIG. 13.—A field of onions being grown for the production of seed.

industry in which one is engaged. Horticultural plants, for convenience in the present discussion, may be grouped roughly into four classes:

1. One class includes plants that are grown for some vegetative part and are propagated commercially by asexual means. The Irish and sweet potato are examples of such plants. Only the plant breeder who aspires to develop a new variety of plant belonging to this class is interested in pollination and fertilization processes. The commercial grower of a named variety of such a crop is not directly or remotely concerned with the processes that result in seed formation.

2. A second class may include plants, grown from seed, of which a vegetative part is used for food or is of other economic value. The seedsman who grows the seed for sale is interested in the processes that precede seed formation, that is, in pollination and fertilization. But the grower of the commercial crop usually harvests the crop before it has opportunity to produce seed. Examples of this class are the leafy vegetables, asparagus, onions, and many others.

3. Aside from their use for propagation, the seeds of many plants are of economic value. They may be used directly as a source of food. This is true of the bean, garden pea, pecan, English walnut, and almond. Seeds may also be the source of

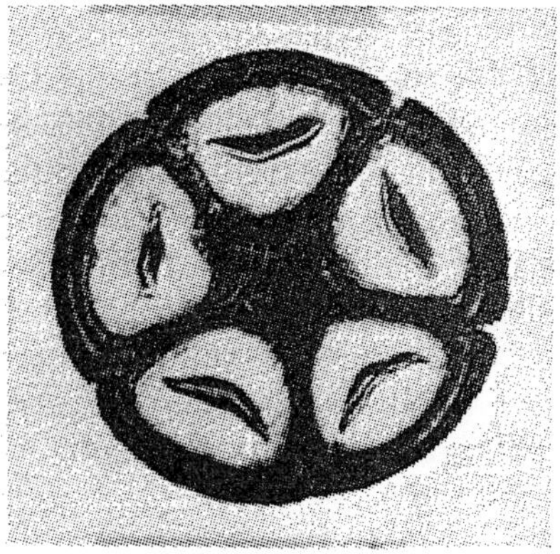

FIG. 14.—Cross section of tung-oil fruit, showing typical arrangement of seeds.

oils or other extracts that are of commercial use; the tung nut, a source of tung oil, is an example of the latter case. Seed development depends on the fertilization process. It is evident, then, that factors and conditions that influence pollination and fertilization are of direct concern to the grower whose primary and chief concern is the production of seeds—whether for use in propagation or as a source of food or some by-product. The seedsman is interested in the source and inherent quality of the pollen, the commercial grower only in effective pollination, regardless of origin of the pollen.

4. A fruit from the botanical viewpoint is the matured ovary. Fruits usually contain seed that, except in cases of parthenocarpy, are essential to the development of the ovary into a fruit. In

the flowers of fruit trees, pollination and the growth of the pollen tube down the style provide the initial stimulation to growth processes that prevent immediate falling off of the flower. In order to keep the fruit developing normally, fertilization and seed formation are necessary in most cases. Seed formation is, therefore, of the greatest importance in obtaining a set of fruit, because without it the young fruits are shed. This is true of the orchard fruits such as the apple, pear, peach, and plum. It applies, likewise, to the vegetable fruits—the cantaloupe, watermelon, pepper, tomato, and others. Fruits that normally contain many seeds are usually lopsided or misshapen if they are not properly pollinated and, hence, do not have the number of seeds characteristic of the species. Apples, for example, normally have from five to ten seeds. Those having only from one to three seeds are not symmetrical. It is thought that the presence of seeds is necessary for the normal development of the conducting tissue in the fruit, which connects the fruit with the tree. It is evident then that the growers of fruits, with certain exceptions, are concerned with effective pollination and fertilization of fruit blossoms. In many cases the seeds are not only valueless but actually objectionable; yet they influence the development of the ovary, which, in the case of fruits, is the final objective.

Factors Influencing Fruit Setting.—With a few exceptions, which will be mentioned later, it is necessary that flowers be fertilized in order to set fruit. There are several factors that may operate to discourage or prevent fruitfulness and are therefore of vital concern to the fruitgrower.

1. *Defective Flower Parts.*—Incomplete development of pistils or stamens may result in sterility. Many species of native plums and other fruits regularly produce certain numbers of defective pistils, which are incapable of being fertilized and drop from the tree at the time of petal fall. The pecan frequently produces some small pistillate flowers, which abcise from the peduncle at an early stage of growth. These defective pistillate parts vary in number from year to year and usually account for a comparatively small percentage of the total drop.

Sterile or aborted pollen is produced by several species of grapes and by many of the citrus fruits. In the former case, the inability of the pollen to effect fertilization will cause failure of the setting of the fruit, unless provision is made for a supply of effective pollen from other sources. In citrus fruits, lack of pollination results in seedlessness, but the fruits will normally continue growth until mature. They are more valuable by virtue of being seedless.

2. *Incompatibility.*—Pollination does not always insure fruit setting. In many fruits incompatibility exists between pollen and pistils of the same plant. Such conditions result in self-sterility. Plants that are propagated by vegetative parts from a self-sterile parent are likewise self-sterile and are intersterile with other plants of the same variety. Instances of self-sterility are not rare in horticultural plants. The J. H. Hale peach is generally self-sterile. An orchard of this variety of trees tends to be unproductive except when proper provision is made for pollination. Most of the important commercial varieties of apples are listed as self-sterile, or partially so. For each important variety that is known to be self-sterile, at least one other variety known to be an effective pollenizer should be planted. Several of the leading pear varieties are regarded as self-sterile under certain conditions; the two leading varieties, Bartlett and Keiffer, are regarded as "commercially self-sterile." Many of the varieties of the Japanese plums, such as Abundance, Wickson, Burbank, and others, are regarded as self-sterile. The McDonald blackberry is self-sterile and, consequently, unfruitful when isolated; the Haupt variety is an effective pollenizer for it.

When incompatibility exists between pollen and pistils of different varieties or species, they are said to be *intersterile.* Each of three leading varieties of sweet cherries (Bing, Lambert, and Napoleon) is self-sterile, and the three are intersterile. Hence, mixed plantings of them will not produce fruit unless the trees are within range of some other variety or varieties that are interfertile with them. Black Tartarian and Black Republican are varieties that are satisfactory as pollenizers for each of the three.

The Bartlett and Seckel varieties of pears are regarded as commercially intersterile. In addition, each variety is self-sterile, and, hence, the two varieties are not fruitful when planted alone, or in combination, without an effective pollenizer.

3. *Dichogamy.*—Plants frequently exhibit a difference in the time of pollen shed and the time of pistil receptivity. The period when pollen is shed may not coincide with the time when the

pistil is receptive. If the two periods are entirely distinct, the condition is known as *complete dichogamy;* if some overlapping occurs, it is *incomplete dichogamy.* For example, a variety with pollen shedding from April 11 to April 18 and pistils becoming receptive from April 20 to April 28 would represent a condition of complete dichogamy. If pollen is shed before pistils are receptive, the plant is said to be *protandrous;* if pistils are receptive first, *protogynous.*

Isolated plants of a dichogamous variety are, normally, unfruitful. Plants that are incompletely dichogamous, if isolated, will set fruit only during the period of overlapping, and the quantity of fruit set is determined, largely, by the length of the period during which pollen shedding and pistil receptivity coincide. It should be kept in mind that dichogamy also results in self- or intersterility, not as a result of incompatibility but merely through the operation of the time factor.

Data obtained in Texas showed dichogamy to be a factor limiting the fruitfulness of pecans. The Burkett, Success, Schley, and other leading varieties have been protogynous in most years. In only a few cases was there sufficient overlapping in blooming to allow self-pollination. The Moore, on the other hand, was found always to be protandrous and produced pollen each year in time to pollinate any variety that has been under observation. At the time that the pistils of Moore became receptive, pollen from other varieties was available.

4. *Dioecious Plants.*—Plants that bear only staminate flowers never produce any fruit, since no ovary capable of developing into a fruit is borne; isolated plants that bear only pistillate flowers are seldom fruitful unless some special provision is made to insure effective pollination. The Smyrna fig, a dioecious species that bears only pistillate flowers, was not fruitful when first introduced into the United States because pollen that was necessary for fruit setting was not provided. In pollination of this fruit, Capri figs containing insects are placed in wire baskets and hung in the Smyrna trees. These are replaced at intervals of 2 or 3 days in order to insure a continuous supply of viable pollen. The Capri figs are grown in protected locations in order to prevent possible loss of the overwintering crop of figs, which contain the fig wasps. A similar practice of planting is followed with the male date palm. With some dioecious species,

such as the Muscadine grape and pistachio, the staminate plants are interplanted to provide proper pollination.

5. *Environmental and Nutritive Factors.*—In addition to the factors mentioned above, fruit trees may fail to set fruit properly due either to frost and other climatic factors or to internal nutritive conditions. The importance of frost and cold weather is often quite apparent in sections of the South where warm winter weather causes early blossoming of peach, plum, and other fruits. In addition to the damage due to outright killing of the pistils, cool weather may slow pollen-tube growth so that the ovule does not become fertilized before an abscission layer forms and causes dropping of the flowers.

Damage to flowers by frost may be lessened by (*a*) selection of varieties that are slow to start growth under the influence of warm winter weather, (*b*) planting on land sloping toward the north, (*c*) providing for proper air drainage of the orchard site, and (*d*) certain cultural practices that delay the beginning of the dormant period in the fall, with the resulting tendency to delay early blossoming. The effect of high humidity on pollen shed, as an environmental factor, has been mentioned previously in the case of the pecan.

Unfavorable nutritive conditions within the tree often result in its failure to set fruit. Partly developed fruits may drop off owing to the plants' inability to furnish the necessary nourishment required in growth of vegetative and flower parts. The nutritive factor may often explain why certain varieties bear a heavy crop one year and a very light crop during the next season. The explanation may be in the fact that undernourished or weak trees often produce defective pistils, fail to develop viable pollen, or fail to develop any flower buds at all. It may determine, also, which of the branches will bear fruit; evidence indicates that those which are more vigorous or are in a more favored position with respect to a nutrient supply are more likely to bear fruit. This is especially noticeable during light-crop years.

Practices That Encourage Fertilization.—It is suggested in previous paragraphs that some plants are not fruitful because viable pollen is lacking. This condition is especially prevalent in fruit orchards. It may be overcome by distributing, through the orchard, blossoming branches of a variety that produces good pollen during the time when pollen is needed. Large branches, from 1½ to 2 inches or more in diameter, are preferred. The pollen of flowers of such branches matures and is available for distribution by wind or insects for several days. The branches are placed in vessels of water and are replaced by fresh ones as the flowers wither. A permanent source of such pollen may be provided by grafting into each tree a scion of a good pollenizer. As the scion grows it becomes a part of the tree and provides pollen for future years. A third way of encouraging pollination is to plant trees of good pollenizers at intervals throughout the orchard. The customary practice is to plant one pollenizer to each ten trees. In many cases, several varieties of commercial value may be interplanted, which will fertilize each other. In a variety intended primarily to provide pollen, fruit production may be of secondary importance, but usually a pollenizer may be found that will produce desirable fruit as well.

Seedless Fruits.—Some plants commonly produce fruits that contain no seeds. The conditions responsible for seedlessness may be conveniently discussed under two headings:

1. *Parthenocarpy.*—Fruits that set and mature without being fertilized are known as *parthenocarpic* fruits. Two different types are recognized: vegetative and stimulative. When plants set and mature fruit without pollination, the condition is known as *vegetative parthenocarpy.* Examples are found in the banana, Japanese persimmon, English varieties of cucumbers, orange, grapefruit, and fig. Many varieties of figs, if pollinated, may develop and mature viable seeds; if not, the ovule wall continues to develop but, at maturity, is empty. In either case, the ovarian tissue and receptacle develop into the edible "fruit." In some cases, the st mulation of pollination on the ovarian tissue is essential, although fertilization of the ovule does not follow. This condition is known as *stimulative parthenocarpy.* Examples are found in the pear, Jerusalem cherry, Thompson seedless grape, and some varieties of squash. Thompson seedless grapes may be rendered seedless, also, by embryo abortion, discussed in a later paragraph.

Fertilization of parthenocarpic fruits might be objectionable by causing the fruit to develop seed instead of being seedless and by causing the mother plant to be less prolific. In the case of cucumber varieties that are capable of maturing fruit parthenocarpically, flowers that are pollinated and fertilized produce seed-bearing fruits which are usually different in shape from seedless fruits. They tend to be angular in cross section instead of cylindrical. Cucumber vines of varieties that produce parthenocarpic fruits are very prolific of seedless fruits; if pollinated the cucumbers have seed and the vines are less prolific.

The Hachiya variety of Japanese persimmon will set fruit readily without pollination, in which event fruits are seedless. In mixed plantings where pollination is likely to occur, seeded fruits are frequently produced. These have black, discolored areas immediately surrounding some of the seeds and are considered to be inferior in quality to the seedless fruits.

2. *Embryo Abortion or Killing.*—Seedlessness may be due to other causes than parthenocarpy. As a result of the fertilization process, an embryo is formed. If its growth is not arrested, it forms a seed. Fruits in which the development of the embryo is prevented by internal or external factors sometimes mature without seeds. Embryos are often killed by cold or other conditions that do not affect the ovary adversely. If the fruit has developed far enough so that its growth will continue without the stimulus of the developing seeds, a seedless fruit is formed. However, it should be stated that, as a rule, death of the embryo results in premature shedding of fruit. If fertilization takes place in Thompson seedless grapes, the fruits when mature are seedless because of embryo abortion. This phenomenon has been observed, also, to result in seedless fruits in plum and cherry, though such fruits have no special merit because of the presence of the bony endocarp.

Polyembryony.—In some species, seedlings that are not the result of fertilization are produced. Fertilization takes place and one embryo that has heritable factors from both parents is produced. In addition to this embryo a number of vegetative embryos that give rise to seedlings identical in heritable qualities to the mother plant are produced from the nucellus of the ovule. This explains why several seedlings frequently develop from one seed. This condition is known as *polyembryony* or *apogamy*. It has recently been shown that at least one species of the apple produces apogamic embryos. Apogamy is not uncommon among plants and has been observed frequently in the production of citrus rootstocks.

Questions

1. What are the three essential parts of a seed? What is the important function of each part?

2. Name the parts of a complete flower. What is a perfect flower? An imperfect flower? A monoecious species? A dioecious species?

3. Define pollination, self-pollination, and cross-pollination.

4. What are the principal agents of pollination? What weather conditions discourage pollination by wind?

5. List the classes of horticultural plants.

6. What is meant by dichogamy?

7. Tell what is meant by polyembryony or apogamy. Of what significance is it?

Suggested References

1. ADRIANCE, GUY W.: Dichogamy in the Pecan, *Proc. Amer. Soc. Hort. Sci.*, **26**: 435–439, 1930.
2. FROST, H. B.: Polyembryony, Heterozygosis, and Chimeras in *Citrus*, *Hilgardia*, **1**: 365–402, 1926.
3. JONES, H. A., and S. L. EMSWELLER: Methods of Breeding Onions, *Hilgardia*, **7**: 625–642, 1933.
4. KNOWLTON, H. E.: Pollination of Fruit Trees in West Virginia, *W. Va. Agr. Exp. Sta. Circ.* 62, 1933.
5. PHILLIPS, E. F.: Honeybees for the Orchard, *N. Y. (Cornell) Agr. Ext. Bull.* 190, 1930.
6. TUKEY, H. B., and K. D. BRASE: Random Notes on Fruit Tree Rootstocks and Plant Propagation, III, *N. Y. (Geneva) Agr. Exp. Sta. Bull.* 682, pp. 29–30, 1938.
7. WEBBER, H. J.: The Economic Importance of Apogamy in Citrus and Mangifera, *Proc. Amer. Soc. Hort. Sci.*, **28**: 57–61, 1932.
8. WOODROOF, J. G.: Studies of the Staminate Inflorescence and Pollen of *Hicoria Pecan*, *Jour. Agr. Research*, **40**: 1059–1104, 1930.

CHAPTER III

FLOWERS AND FRUITS

Although flowers of many plants are borne singly on a stalk or stem, in numerous other cases the flowers are borne in clusters known as *inflorescences.* The stems that bear the individual

FIG. 15.—Flowers of the peach. Note that they are borne singly or in pairs.

flowers of the inflorescence are known as *pedicels,* and the main stem of the inflorescence from which the pedicels arise is called the *peduncle.* The stem of a flower borne singly is also known as the peduncle. A flower attached to the main axis or peduncle, without any stalk or pedicel, is known as *sessile*. The end of the pedicel is generally flattened into a structure known as the *receptacle*, to which the flower parts are attached.

TYPES OF INFLORESCENCES

There are two distinct types of inflorescences, the *indeterminate* or *racemose*, and the *determinate* or *cymose*.

FIG. 16.—Pistillate and staminate inflorescence of walnut. Catkins borne singly from secondary buds on 1-year wood; pistillate flowers in spike terminating new growth.

Racemose Inflorescences.—The indeterminate or racemose inflorescence is produced on an axis that continues to grow and produce flowers laterally. Naturally, the older flowers are at the base of the inflorescence, and the youngest near the growing point.

1. *Spike.*—In this type of inflorescence the main axis is elongated and the flowers are sessile. The inflorescences of many plants are of this type. Examples are gladiolus, snapdragon, and the pistillate flower of pecan.

The *catkin* is a spike that bears only pistillate or staminate flowers and falls off the plant. The peduncle of the catkin is

FIG. 17.—Flowers of pecan. Three-stalked catkins borne in pairs from buds on 1-year wood; pistillate spike borne terminally on new growth.

flexible. Examples of plants that bear catkins are the pecan, willow, oak, and walnut.

2. *Raceme.*—In this type the main axis of the inflorescence is elongated and flowers are borne on pedicels of about equal length at maturity, with the result that a single raceme tends to become more or less cylindrical in form. The inflorescences of the hyacinth, cabbage, and yucca are examples.

3. *Corymb.*—A corymb is an inflorescence in which the main axis is elongated but the pedicels of the older flowers are longer

FIG. 18.—Inflorescence of the pear, a corymb.

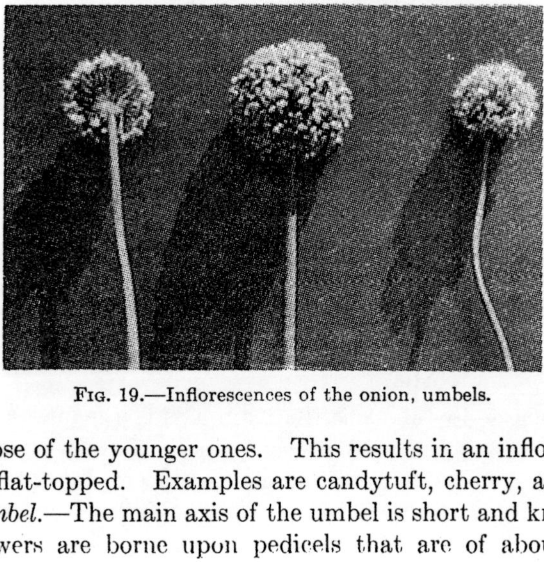

FIG. 19.—Inflorescences of the onion, umbels.

than those of the younger ones. This results in an inflorescence that is flat-topped. Examples are candytuft, cherry, and pear.

4. *Umbel.*—The main axis of the umbel is short and knob-like. The flowers are borne upon pedicels that are of about equal

length. The pedicels are attached upon the top of the peduncle at about the same point. Examples of the umbel are onion and carrot.

5. *Head.*—The axis of the head is very short, and the flowers are sessile. In many instances the entire head is regarded as the flower, when, in reality, it is an inflorescence that contains many, perhaps hundreds, of separate flowers. Sunflower, lettuce, and goldenrod are examples of this type.

FIG. 20.—Fig tree, showing fruit formation in axils of leaves on shoots of the current season.

Cymose Inflorescences.—In the determinate or cymose inflorescence, the main tip ceases to grow, owing to the formation of a terminal flower. The other flowers are borne on lateral branches farther down the axis, and the youngest of the flowers in the cluster are found at the greatest distance from the tip. The growth and elongation of the floral axis is definitely terminated by the formation of the flower at the tip or apex of the peduncle. It will be noted that, in the cyme, the youngest, and last, flower to bloom is farthest from the tip, whereas in the raceme the opposite condition is true.

1. *Solitary Flower.*—Many plants produce a solitary flower that is terminal on the peduncle. Examples of such plants are

the rose, tulip, and poppy. These represent the simplest type of cymose inflorescence.

2. *Fascicle.*—An inflorescence of the cymose type with the flowers very closely crowded is known as a *fascicle.* The phlox is a plant that bears an inflorescence of this kind, and that of the apple is very similar.

Fig. 21.—Cluster of 56 pecans which developed apparently from several combined inflorescences.

3. *Glomerule.*—Cymes in which the flowers are more condensed than in the fascicle are known as *glomerules.* They can be distinguished from a head by the order of opening of the flower buds, which is from the center progressively outward. Certain species of dogwood illustrate this type of flower cluster.

Compound Inflorescences.—In some inflorescences the peduncle is branched, and from it clusters of flowers arise instead of solitary flowers. This may be true in flower clusters of either the determinate or the indeterminate type. In some flower clusters, both simple and compound types are represented. This is true in the grape and in *Ligustrum.* There are several different kinds of compound inflorescences. The simplest type is the *panicle,* in which a group of flowers much like an ordinary raceme is borne with the same relationship to the central axis as that between the pediceled flower and the peduncle of the simple raceme. Possibly the most extreme form of the panicle is illustrated by the oat, in which several orders of branches may arise successively. In the inflorescence of certain plants, such as the parsnip, several umbels occur on the central axis at points where single flowers occur in the simple inflorescence. Such a cluster is known as a *compound umbel.*

TYPES OF FRUITS AND SEEDS

From the botanical viewpoint, the fruit is the matured ovary with its seeds and other parts of the flower that are intimately

FIG. 22.—Cross and longitudinal sections of tomato, a berry.

associated with it at maturity. The seed is the matured ovule and is enclosed in the fruit. Most fruits consist of a single ovary, but there are a number that include several ovaries, and some that consist of ovaries of several flowers coalesced into a unified mass.

Simple Fruits.—These fruits are composed of a single enlarged ovary to which other parts may or may not be attached. They are of two general types: with *fleshy* pericarp and with *dry* pericarp.

1. *Fleshy Fruits.*—A *berry* is a fruit with one or more seeds wholly or partly embedded in a fleshy endocarp and mesocarp.

The date, tomato, and grape are examples of true berries. The various citrus fruits are of a berry-like type, known as *hesperidium.* These have a thick, leathery rind or pericarp and a juicy pulp enclosed in partitions. The *pepo* is also a berry-like fruit, the outer part of which is receptacle tissue attached to the exocarp. The flesh is principally mesocarp and endocarp. This type

FIG. 23.—Cross sections of orange, a type of berry known as hesperidium.

characterizes the fruits of Cucurbitaceae, such as watermelon, squash, pumpkin, cantaloupe, and cucumber.

A *drupe* is a fruit consisting of a single seed surrounded by a hard or bony endocarp, a fleshy mesocarp, and a thin exocarp or skin. Examples are peach, plum, nectarine, cherry, apricot, almond, and olive. In the almond, the mesocarp and exocarp dry up and split away at maturity.

A *pome* is a fleshy, many-seeded fruit in which the flesh is very largely made up of receptacle tissue surrounding the ovary and

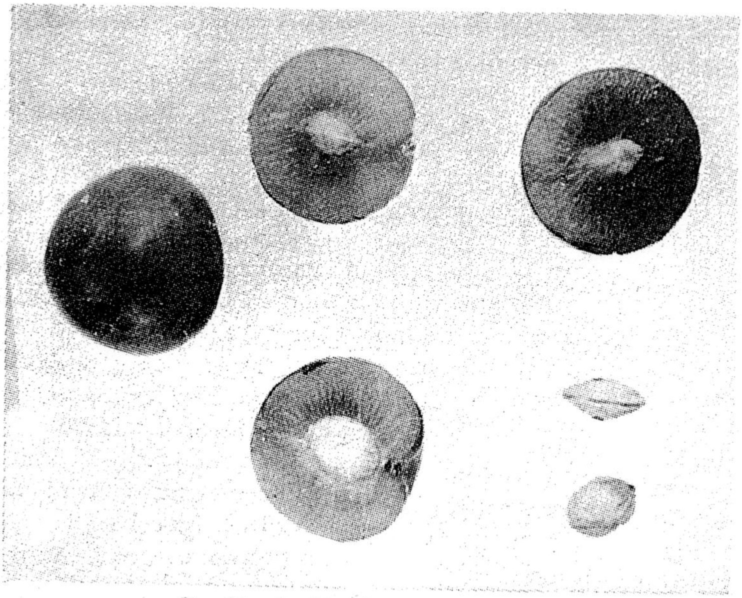

FIG. 24.—Sections of the plum, a drupe.

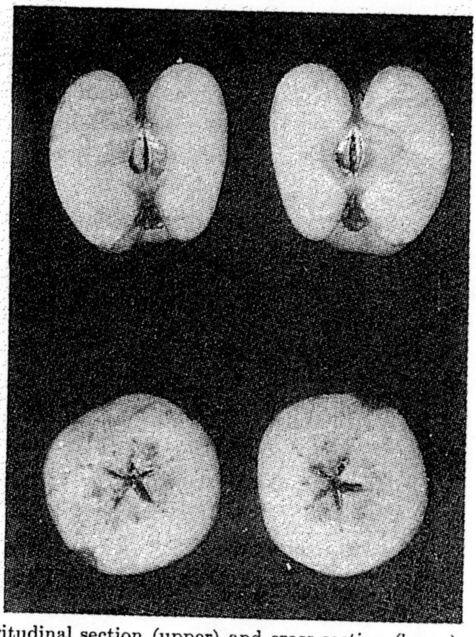

FIG. 25.—Longitudinal section (upper) and cross section (lower) of the apple, a pome.

the seed are enclosed in bony or leathery carpels. Examples are apple, pear, quince, May haw, and loquat.

2. *Dry Fruits.*—The dry fruits are further separated into *dehiscent* and *indehiscent*, the former splitting open when ripe and the latter not splitting. The most important of the types of dry dehiscent fruits follow:

A *legume* is a dry fruit composed of a single carpel and splitting along both sutures at maturity. The ventral suture is the line of fusion of the carpel and the point of attachment of the seed, while the dorsal suture corresponds to the midrib of the carpel. Species of the Leguminosae produce fruit of this type.

A *follicle* is a dry fruit made up also of one carpel, which at maturity dehisces along the ventral suture only. Examples are milkweed, larkspur, columbine, peony, and marsh marigold.

A *capsule* is a dry fruit made up of two or more carpels. This type of fruit may dehisce along the line of union (ventral margin), along the middle of the carpel (dorsal wall), or by pores.

A *silique* is a dry fruit derived from a superior ovary consisting of two carpels. It has two parietal placentae, connected by a false partition, which separates the cavity of the fruit into two compartments. At maturity the carpels separate from the false partition and leave their margins and the placentae attached to the false partition, which, in turn, remains attached to the torus. This fruit is characteristic of the mustard family, including cabbage, turnip, rape, mustard, and radish, as well as shepherd's-purse.

The most important of the dry indehiscent fruits follow:

An *achene* is a one-seeded dry indehiscent fruit in which the seed is attached to the ovary wall at one point only. The buckwheat fruit is a good example, the hull consisting of the pericarp and parts of the calyx. Other examples are sunflower, buttercup, and "seed" of the strawberry.

A *caryopsis* is a dry indehiscent fruit in which the seed coats have so completely fused with the pericarp that they can no longer be distinguished as separate organs. The so-called "seed" of corn and small grains are really fruits in which the pericarp is united with the parts it encloses.

A *samara* is an indehiscent fruit in which the ovary wall is expanded to form wings, which aid the seed in its dissemination.

The ovary is sometimes two-celled, as in maple. Other examples are ash and elm.

A *schizocarp* is a dry fruit of two carpels, or a double achene, attached by the inner faces to the slender stalk or carpophore, from which it separates at maturity. Each half of the double achene is called a *mericarp*. Between the ridges on the back of the mericarps are the oil tubes or vittae, to which the aromatic flavor of the fruit is due. This fruit is typified by the carrot family.

The *nut* is a hard, one-seeded fruit, resulting from entire ovaries, the walls of which become hard throughout but do not dehisce. Examples are acorn, chestnut, and hazelnut. The almond is the stone of a drupe. In the pecan, however, the fruit is a true nut, and the husk is derived from the calyx lobes instead of the outer ovary wall.

Aggregate Fruits.—An aggregate fruit is derived from a single flower having a large number of pistils. The structure of such a fruit is that of a single receptacle upon which are massed a large number of so-called "fruitlets," each of which is like a small fruit. In the dewberry, blackberry, and raspberry, the individual fruits are small drupes; in the dewberry and blackberry these drupelets adhere to the receptacle when it is detached from the stem, while in the raspberry, the receptacle adheres to the stem and the fruit separates as a hollow cup. In the strawberry, the small fruits are achenes or dry fruits, and the receptacle is the edible portion.

Multiple Fruits.—The multiple fruit is developed from the ovaries of many separate and yet closely clustered flowers. The most familiar of these fruits are the mulberry, pineapple, and fig. The individual flowers in the pistillate inflorescence of the mulberry are crowded together closely on the axis. Each flower possesses a single, one-celled ovary, which develops into a nutlet enclosed by the thickened, juicy calyx lobes. These separate fruits become crowded together and coalesce, to form the mulberry fruit.

The flowers of the fig are borne on the inner wall of an enlarged, fleshy hollow receptacle, both staminate and pistillate flowers usually occurring. The pistillate flowers have single, one-celled ovaries, developing into nutlets, which are imbedded in the succulent flesh on the inside of the receptacle. This type of fruit is sometimes called a *syconium.*

The pineapple has an elongated central axis on which are borne numerous sessile flowers. The fleshy bases and ovaries of these flowers are fused with the sepals to form the edible part of the pineapple.

Accessory Fruits.—Those fruits in which a major part of the ripened fruit has developed from other parts than the ovary and ovule are known as *accessory fruits.* The apple is a simple fruit but also accessory because a large part of the flesh is derived from the receptacle. In the strawberry the receptacle comprises practically the entire edible portion and in the blackberry and

FIG. 26.—External views and cross section of fig, a multiple fruit.

dewberry a part of it. They are hence accessory fruits. The mulberry is another such fruit, consisting in part of the main axis and fleshy bracts. The fig fruit consists partly of receptacle and is also an accessory fruit.

PLANT DISSEMINATION

Many plants have structural modifications of fruits, seeds, or other parts that facilitate or encourage dispersal. The extent to which plants possess such modifications determines their mobility. Those that have large heavy fruits or seeds are relatively immobile; those which produce light fruits or seeds, equipped perhaps with wings or similar appendages, have high mobility. Dissemination, or migration, depends upon the mobility of plants and also upon various agencies or factors that tend to move plants from one place to another. Those species which, under natural conditions, have the best facilities for distribution are likely to be found over the greatest range of territory. Oceans, deserts, and mountain ranges act as natural barriers to prevent plant migration by natural means. As a general rule the distribution of a species is also dependent upon its age. The pecan is undoubtedly of comparatively recent origin, as shown by its limited distribution. It is not to be supposed that the small area of this country in which it grows wild is the only region to which it is adapted.

Plant Introduction.—Man either purposely or by chance is often responsible for the distribution of plants. The American colonists very early introduced into America plants that had been grown in the countries from which they came. The United States Department of Agriculture maintains plant explorers in several foreign countries, looking both for new fruits and for more desirable forms of those which are now available. Valuable horticultural plants have been introduced by these explorers, many of them coming from China and South and Central America. Material secured is tested in government introduction gardens before being distributed widely. Some of the most useful fruit and vegetable plants are native to America, the principal ones being the tomato, potato, corn, bean, strawberry, grapes, blueberry, cranberry, plum, crab apple, and pecan. Several others, however, such as the cultivated apple, pear, peach, cherry, citrus fruits, date, fig, and most species of vegetables, have been introduced from foreign countries. Introductions are frequently valuable in themselves, but more often their chief value is for breeding work.

Wind.—Wind is probably the most effective natural agency in the dissemination of plants. The seeds and fruits of many plants have special structures and modifications that enable the wind to carry them. The fruits of maple and elm have wings, and those of the lettuce have hairy disseminules that enable the seeds to be borne by the wind. In other plants the seed is modified so as to be easily blown about. Examples of these are the willow, milkweed, cotton, and cattail. Tumbleweeds, which roll across the prairies as whole plants, illustrate a special case of wind dissemination. The Russian thistle is a plant of this kind. The plant breaks off at the ground after frost and may be blown a great distance, frequently 50 or 75 miles, during the winter months. The small seeds are so enclosed that they do not all fall out, but instead are scattered along the path of the plant as it is swept along by the wind.

Water.—Ocean currents, running streams, and lakes are means whereby plants are scattered. Seeds and fruits have various modifications that enable them to float. The thick, fibrous husk and the hard shell of the coconut are an illustration. Many nuts, such as the pecan and walnut, are readily distributed by water. Seeds of grasses and noxious weeds are often brought in from points upstream by overflows of rivers or creeks.

Animals.—Seed may be distributed by animals in three principal ways: by carrying them in fur, hair, or feathers; by eating the seeds and passing them through their digestive systems; or by storing or burying them. Cockleburs, clover, and needle grass readily become attached to the hair or hide of animals. The fruit of *Martynia*, commonly known as devil's-claw, splits in a way that forms two incurving hooks, which readily attach themselves to animals. The seeds of many edible fruits, such as peach, plum, and berries are carried in the digestive system of certain animals. Nuts are frequently transported by squirrels, blue jays, and crows and dropped at a considerable distance from their point of origin.

In a few cases, plants are disseminated by parts other than seed. This is true of Bermuda grass, which is easily carried from one place to another attached to the feet of animals.

Propulsion.—The fruits of some plants are characterized by an explosive action when ripe, and the seeds are thereby scattered. The distance to which they are propelled may not be great, and the rate of distribution is slow, but when continued indefinitely the effect becomes of importance. Propulsion of seeds is due to the unequal drying of different layers of the ovary wall, resulting in strains that are suddenly overcome when the fruit explodes. The fruits of bull nettle, violet, "sandbox," castor bean, and cowpea illustrate this phenomenon.

Questions

1. What is an inflorescence?

2. What is the essential difference between a determinate and an indeterminate type of inflorescence? Between a spike and a raceme? Distinguish between sessile and pediceled flowers.

3. What is the distinction between a fruit and a seed? Between simple, aggregate, and multiple fruits?

4. Distinguish between fruits with fleshy pericarps and those with dry pericarps. Give examples.

5. Give examples of accessory fruits. What tissues comprise the edible part of each?

6. List *fruits* that have special structures to facilitate dispersion. List *seeds* that have such special structures.

7. What are the principal agencies for the dissemination of plants?

Suggested References

1. Fiske, Jessie G.: Fruits and Seeds of Some Common New Jersey Trees, *N. J. Agr. Exp. Sta. Circ.* 355, 1936.
2. Gardner, V. R., F. C. Bradford, and H. D. Hooker: "Fundamentals of Fruit Production," McGraw-Hill Book Company, Inc., 1939.
3. Pool, Raymond J.: "Flowers and Flowering Plants," McGraw-Hill Book Company, Inc., 1929.

CHAPTER IV

GERMINATION OF SEEDS

The embryo formed as the result of fertilization normally makes only a limited amount of growth before its development is checked. It becomes dormant as the ovule ripens into a seed and remains in a condition of arrested development within the mature seed until it is subjected to conditions favorable to growth. This resumption of growth is known as *germination*, and the process is said to be complete when the radicle and plumule break through the seed coat.

THE PROCESS OF SEED GERMINATION

Water Absorption.—For the initiation of germination it is essential that the seed be subjected to a favorable supply of moisture. With most seeds, it is sufficient to place them in a soil or other medium with a suitable moisture content. Some seed, however, which have impervious seed coats, are unable to take up sufficient water for germination without the aid of special treatments.

The seed coats absorb water and become soft, as the water penetrates the seed, and the enclosed parts, including the endosperm and embryo, swell and push the seed coats off. This increased water content of the seed is also a necessary condition for the processes of gas interchange, chiefly the intake of oxygen and the giving off of carbon dioxide. Other processes that go on with the increased water content of the seed are the movement of soluble food and the resumption of activity in the protoplasm.

Movement and Utilization of Foods.—The seed contains relatively large quantities of stored food, in such insoluble forms as starch, fat, and protein. In order for these materials to be used by the seed, they must be transformed into soluble compounds. Certain specific enzymes, produced within the cells, carry on this process of digestion, which renders the stored foods available.

Following digestion there is a movement of food from the storage cells to the growing parts of the germinating seed, which are principally the growing points of the radicle and the plumule. The foods are used up in various processes at the growing points, while continued supplies are made available in the storage cells; as a result, the movement continues as a process of diffusion.

At the tips of the radicle and the plumule, new cells and tissues are being formed, and elongation of others is in progress. The food supply is utilized to form new protoplasm, to thicken the walls of new cells, and to form wood fibers, conducting tissues, and various other specialized products and structures associated with growth.

In the synthesis of organic compounds in the leaf, the kinetic energy derived from the sun was converted into potential energy. Energy of growth is provided in the breaking down of these carbohydrates and other stored foods. This decomposition of stored foods occurs in the process of respiration, which releases water and carbon dioxide, in addition to energy. Only a part of the energy so produced is used in production of new tissues, and a considerable portion of it is given off as heat.

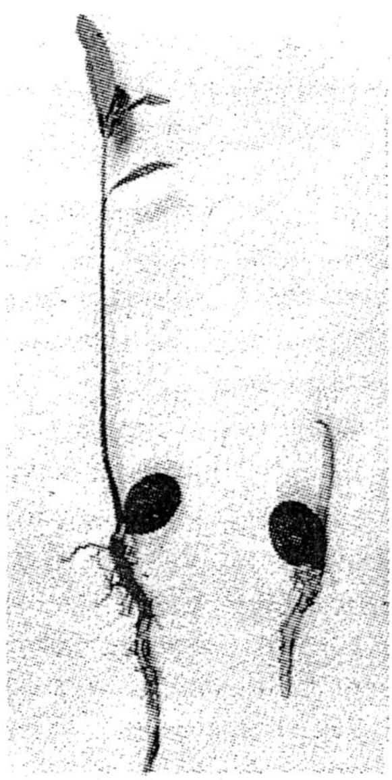

FIG. 27.—Two stages in the early growth of pecan seedlings.

Seedlings.—The process of germination is considered as complete when the seed coat is broken and the radicle and plumule pass outward from the seed, from which time the young plant is spoken of as a *seedling*. This stage continues until it develops into a plant capable of manufacturing its own food and is not dependent upon the food stored in the seed. The plumule produces the part of the plant above the ground, and the radicle the portion below the ground.

Seedlings of ordinary flowering plants are either monocotyledonous or dicotyledonous. The monocotyledonous seedlings produce a temporary root system that is followed by the permanent system, but in the dicotyledonous seedlings the first root system is usually the permanent one. The monocotyledonous seedlings are of two types: in one of these, as typified by corn, the cotyledon remains in the ground; in the other, as typified by the onion, the cotyledon is raised to the surface. The same difference occurs in the seedlings of the dicotyledonous plants. In melons, radish, beans, and other seedlings of this type, the cotyledons are pushed out of the ground by elongation of the hypocotyl and often function as leaves for some time. On the other hand, the plumules of seedlings of the garden pea push upward through elongation of the stem of the epicotyl, while the cotyledons remain in the ground. It is generally considered that seedlings of this type can emerge from the soil with less difficulty than those which must raise the cotyledons.

REQUIREMENTS FOR GERMINATION

In order for seed to germinate they must have proper conditions with regard to moisture, temperature, and oxygen, and seeds must be viable to be able to respond to these conditions. Any one of these four may become a limiting factor in preventing germination.

Moisture.—The amount of moisture required for germination is usually that which will completely saturate and soften the seeds. The water absorbed saturates the cell walls and starch grains, and fills the living cells of the embryo and all empty spaces that exist in the seed. The amount of water required for saturation varies for different seeds. Corn is saturated by 43 per cent of its weight of water; peas require 107 per cent; and sugar beets, 120 per cent. This intake of water is accompanied by a great increase in volume of the seed.

Temperature.—There are wide differences in seeds with regard to the temperature required for germination. Seeds of the so-called "cool-season crops," such as garden peas, cabbage, and carrots, will germinate fairly readily at temperatures as low as 45°F. The optimum temperature for these crops probably ranges between 70 and 80°F. Cool-season crops will germinate at fairly low temperatures and, furthermore, will not germinate if the temperature is too high. Difficulty is experienced in getting carrot seed to germinate in late summer and early fall in the South, on account of the high temperatures. The warm-season crops require higher temperatures for germination. Sixty degrees Fahrenheit is about the minimum temperature for muskmelons and other cucurbits, and they germinate better within a shorter period of time at a much higher temperature. Pepper seed apparently germinate within the shortest time if the temperature ranges between 90 and 100°F. Temperature influences germination through the rate of water intake as well as by increasing the velocity of the processes within the seed.

Oxygen.—The seed is a living structure and requires oxygen. The embryo requires oxygen for the initiation of growth. Under ordinary conditions the seed does not suffer for lack of this element, since the atmosphere contains an abundance. In compact, poorly prepared, or excessively wet seedbeds, germination may be retarded or prevented by a lack of oxygen. The fact that oxygen is required for germination and that water in excess is not detrimental has been proved with asparagus. Seeds in unaerated water showed no germination at all in 2 months, though others, in water through which air was bubbled, germinated in 1 week.

Viability.—A seed is viable if it is capable of germinating. Those that are viable vary widely in vitality, being influenced primarily by:

1. *Conditions of Growth.*—Seed from weak plants are apt to be deficient in stored food and also to have small embryos. They will produce less vigorous plants than those from normal seeds. Atmospheric humidity and temperature during seed development may also influence vitality. Dry atmosphere and absence of very low temperatures are favorable conditions. It is important that seed for planting mature normally. Seeds that are not mature may germinate, but they are less likely to do so under unfavorable conditions. They also lose their vitality more quickly than do seeds that are fully mature.

2. *Age of Seed.*—The vitality of seeds of the different species is influenced in varying degrees by age. Seeds of the willow will lose their vitality in a few days, and those of many tropical plants are said to remain viable only a short time. Among the common vegetable and flower seeds, there is a great variation in this respect. Cucumber, endive, celery, and chicory are said to give satisfactory germination after 8 to 10 years, but dandelion, martynia, onion, parsnip, and sweet corn will last only 1 to 2 years, ordinarily.

In experimental work of the United States Department of Agriculture begun in 1902, sample seeds of 107 species of plants were buried in pots at depths of 8, 22, and 42 inches. Germination tests were made at intervals, and at the end of 20 years 51 species showed capacity for germination. Most of these were weeds; the seed of the cultivated plants, such as grains and legumes, failed to survive.

3. *Storage of Seed.*—The conditions under which seed are stored should be determined primarily by the type of seed and its response to differences in temperature, humidity, and oxygen. A low temperature is of value in preventing early germination of some seeds and in preventing oily seeds, such as pecans, from becoming rancid. Oxygen has a tendency to speed up respiration and other life processes, and may thus cause early deterioration of stored seed.

Many vegetable and flower seeds can be kept in storage for considerable periods of time, provided the one important factor of atmospheric humidity is kept sufficiently low. A combination of dryness and cool temperature is most desirable, but the temperature may vary considerably without causing the seed to deteriorate. In semitropical and tropical countries where the humidity is high it is quite difficult to carry seeds over from one year to another. Most seeds secured for spring planting in the tropics are unsatisfactory for planting the following fall. In a comparison of storage methods for seeds in Puerto Rico it was found that vegetable seeds gave much better results when placed in cloth bags and packed in closed glass jars, with calcium chloride in the bottom as a drying agent. Bean seed remained viable for 2 years in this dryer, while radish, tomato, and beet tested 84 per cent germination, or better, after 5 years. All seed kept in the open had lost viability at the end of 1 year. None of the lots were stored in vacuum or in absence of oxygen.

In this country, work done by the Bureau of Plant Industry showed that vegetable seed remained viable for a much longer time at Ann Arbor, Mich., than at Mobile, Ala. There was a close correlation between rainfall and deterioration of the seed, with higher temperatures contributing as a secondary factor in loss of vitality. In Texas, onion seed that were stored in cellophane bags and in cloth bags at room temperature lost their viability in 2 years, while those that were sealed in glass jars, either at normal pressure or in vacuum, remained perfectly viable for 4 years.

The seeds of most of the fleshy fruits germinate more satisfactorily if they are not allowed to dry out. Many of these seed, however, are handled in the dry condition for convenience and for economy in transportation. Citrus seed give best results when planted immediately upon removal from the fruit. For storage, they should be placed in a moist medium and kept cool. For shipment, they may be dried on the surface and packed in tight containers. Apple and pear seed should be stored under cool, moist conditions as soon as extracted from the fruit. It is a common practice to place them in loosely filled cotton bags and store between cakes of ice. Seeds of peach and related species are stored in moist conditions, a common practice being

FIG. 28.—The onion seedlings in the flat on the right grew from seed which had been kept sealed in air-tight containers for 3 years. Onion seed stored during the same period in cloth bags failed to germinate when planted in the flat on the left.

to pack them in moist sand or soil. Nuts and other oily seeds are commonly stored at very low temperatures in order to preserve them in a viable condition, because of their high oil content and the tendency to become rancid.

Seeds of many plants, especially the grains and legumes, are subject to damage by weevils. They should be stored in tight containers and fumigated at the beginning of the storage period.

DELAYED GERMINATION—CAUSES AND TREATMENTS

In the preceding discussion, it is presumed that seeds will germinate if viable and if subjected to favorable environmental conditions. Most seed will do that. It happens, however, that seeds may be viable and still not respond immediately—nor ultimately in some cases—unless subjected to the influence of certain treatments that modify them in some way and enable them to germinate. Some of the factors and conditions that tend to delay germination are discussed in the following paragraphs.

The Seed Covering.—The seed covering is frequently a factor in retarding the intake of water that is essential for the expansion and growth of an embryo. It is not sufficient that moisture surround the seed; it must penetrate the covering and be absorbed by the embryo before germination can begin. Seeds of alfalfa, olive, pecan, and walnut have *hard* coverings that restrict the rate of absorption of water and, in doing so, delay germination. This is especially likely to be true if the seeds have been allowed to become excessively dry. It should be remembered that the seed covering may comprise other structures than the testa. The shell of the pecan, for example, develops from the ovary wall and is, hence, the pericarp. The bony seed covering of the peach develops from the endocarp of the ovary wall. In others, the bony seed covering is the testa. This is true of the date and persimmon. Some very strong seed coverings, as in the peach, may actually prevent emergence of the growing points unless the outer covering is cracked.

In some cases tissues that surround the seed may delay germination by retarding the passage of oxygen into the region of the embryo. This is thought to be the cause of delayed germination of the cocklebur. It accounts also for the poor germination of freshly harvested lettuce seed of certain varieties. At 30°C. most varieties of lettuce seed will not germinate. Evidence is available that indicates that the *integumentary membrane* and *endosperm* that surround the embryo prohibit the free diffusion of oxygen inward and carbon dioxide outward. The oxygen requirements increase rapidly with an increase in temperature, so that an adequate supply fails to diffuse through the membrane at the higher temperature. A higher temperature is required to inhibit the germination of old seed than for freshly harvested seed of the same variety, which suggests that with age the tissues that inhibit gas exchange become more permeable. If the integument and endosperm are removed, the embryo will germinate at high temperatures. The seed will germinate at the high temperatures without removal of the surrounding tissues if the oxygen pressure is increased. It is known, also, that there is a peculiar relation between germination of lettuce seed and exposure to light, the light requirement being greatest in freshly harvested seed.

Spinach seed does not germinate well at high temperatures. It has been suggested that products of metabolism that arise and accumulate only at higher temperatures are responsible for this failure to germinate. A discussion of some of the treatments used for seeds with strong or impervious coverings follows:

1. *Mechanical Treatments.*—Hard seed coats, regardless of their structure, may be rendered more pervious to water by certain mechanical treatments, such as scratching, cracking, or clipping

FIG. 29.—Barrel "seed scarifier."

them. Seeds treated in a barrel "seed scarifier" germinate better because of the action of an abrasive on the seed coat. Clipping the apical end permits ready absorption of moisture and quick germination of some seeds. This is used effectively on olive seeds, which germinate poorly unless subjected to some treatment that will render the seed more permeable to water.

2. *Chemical Methods.*—Chemicals may also be used to break down the seed covering and render it more pervious to water. Sulphuric acid, used in strengths varying from concentrated to

dilutions of 50 and 25 per cent, has been found effective. Weaker acids, notably acetic, and certain bases have been used also, with varying results. The seed are submerged in a solution of the chemical for a specified time, which is determined by the strength of the solution and the nature of the seed coat. It is important that the seed be washed in water following chemical treatment.

3. *Stratification.*—Another treatment that influences germination by its effect on the seed coat is stratification. The actual method of stratifying seeds may be varied in several ways. The seed may be placed in layers, alternating with layers of sand, in a large box that is left open at the top for the addition of more water as needed, or they may be placed in a shallow pit or trench and covered with earth. The one precaution necessary is to place them where there will be sufficient drainage to keep the soil from becoming water-logged. The soil or sand should be moist but not saturated if the seed is to remain in it for an extended period of time. Stratification keeps the seed covering moist, which permits readier absorption of moisture when the seeds are planted. Stratification is one way of preserving seeds that lose their viability if allowed to remain dry for a considerable time. Seeds of apple, pear, and cherry belong in this class. Afterripening processes, discussed in a subsequent paragraph, may also take place while seed are stratified, if the temperature is favorable.

4. *Soaking.*—Soaking seeds in water for 12 to 24 hours before planting will, in many cases, hasten germination. This would be expected, since water intake is the first step in the process. This soaking does not eliminate the necessity for proper soil moisture, which must still be present to enable the seeds to complete germination. Although unheated water is most commonly used, warm water is absorbed more rapidly. Asparagus seed soaked in water at 30°C. took up their maximum amount of water (about 43 per cent) in 30 days less than other seeds of the same kind in water at 18°C. Some seeds with very hard seed coats may be placed in hot water for a few minutes with good results, but this treatment would be too severe in most cases.

Afterripening Phenomena.—Seeds of some species of plants will not germinate until they have undergone changes that, for convenience, are known as *afterripening processes.*

1. *The Rest Period.*—The time during which mature seeds will not germinate but will retain their viability is known as the *rest period.*

Peach seed of kinds commonly planted for the growing of stocks will not ordinarily germinate until they have undergone a rest period of about 3 months under conditions that provide both a low temperature and moisture. In actual practice, it is sufficient to stratify the seed in moist sand, and hold them at a temperature of 35 to 40°F. Where winters are mild, it is advisable to place the stratified seed in cold storage in order to provide this favorable temperature. Where winters are severe, outside temperatures may be effective, though extremely low temperatures are to be avoided. Ether and chloroform have been observed to hasten germination of some seeds, and it has been suggested that the influence is upon the rest period.

Seeds of apple, pear, some species of roses, and numerous other plants require a rest period before they will germinate. Not all seeds, however, are characterized so definitely by this phenomenon. The pecan, for example, does not have a rest period, and under the influence of excessively high humidity it frequently germinates on the tree before harvest. The same is true of seeds of some varieties of grapefruit.

2. *Immature Embryos.*—The embryos of seeds of certain species of plants are not mature at the time the seeds are apparently ripe. This is true of some species of the buttercup and is also thought to be the reason for delayed germination of carrot and of certain species of *Ilex.* Apparently, little can be done to hasten the maturity of seeds that have such immature embryos. It is important to hold them under conditions of temperature and humidity that will help preserve their viability until such a time that they will germinate.

Vernalization.—Work that was initiated in Russia several years ago by Lysenko indicates that certain seed treatments may influence the plant in its later development. The specific objective is to shorten the period of vegetative growth, and increase yields of seed and plant material. The treatment consists in starting the process of germination by adding a limited amount of water to the seed, and then restricting the growth of the seed by the use of low temperature, drying, or darkness. Work in this country has shown that favorable responses were obtained with white lupine, crimson clover, hairy vetch, and Austrian winter field pea, vernalized and kept for 40 days at 0°C. Seed that were once started into germination and then dried, initiated growth sooner than seed not so treated, and the resulting plants had a shorter vegetative period.

SEED TREATMENTS TO CONTROL DISEASE

In the following discussion, the term *seed* is used to designate a matured ovule, a structure that contains an embryonic plant. In some cases they may be *true seeds*, in others, they may be *fruits* in which the seeds are included as, for example, those of the beet, or lettuce. Tubers, fleshy roots, bulbs, and other vegetative structures that may be used for reproduction and are, hence, sometimes referred to as "seed" will be considered in another chapter.

Many diseases are transmitted by the seed, the organism being borne either upon the surface or within the seed coat. Seed-borne diseases may attack the very young seedling plants either before or shortly after they emerge and prevent a good stand; in other cases the damage may be delayed until the plant approaches maturity.

Diseases Borne on the Seed Coat.—Organisms that produce black rot of cabbage, smut of sweet corn, anthracnose of watermelons or muskmelons, and the fungi that cause "damping-off" of many kinds of seedlings are examples of diseases that may be borne on the seed coat. There are many others that might be mentioned. The seeds may become infected in several ways; probably the most common source is diseased mother plants. Spores of the fungi cling to the seed coat or are imbedded in it. When the seed germinates, the fungus spore also germinates. In the case of damping-off the disease shows up at once. In others, anthracnose of watermelon for example, the presence of the disease may not be apparent until the plants have made considerable growth. These diseases and others may be disseminated in other ways than on the seed. Seed treatments to control them, however, are considered to be worth-while precautionary measures. Suggestions for the use of some of the more common disinfectants are given in the following discussion:

1. *Bichloride of Mercury.*—This chemical is considered to be one of the most effective disinfectants for seed. The solution usually recommended is made by adding 1 gram of the bichloride to 1,000 cubic centimeters of water. A solution of the same strength may be prepared in greater volume by dissolving 1 ounce of the bichloride crystals in 8 gallons of water. Soaking the seed for 15 minutes is recommended to control diseases of cabbage and related plants, 10 minutes for diseases of cantaloupe, watermelon, squash, and cucumbers; and 8 minutes for tomato, pepper, and eggplant. After the treatment is completed the seed should be rinsed several times in water. Metal containers should not be used for the treatment; glass, stoneware, or wooden vessels may be used.

2. *Copper Sulphate.*—Soaking seed in a solution of this chemical is effective in destroying certain seed-borne diseases. It has been recommended particularly for destroying the damping-off organism on spinach and tomato seed. The solution is prepared by adding 1 to 2 ounces of the copper sulphate to 1 gallon of water. The seed are soaked for approximately 1 hour, after which they are dried without rinsing preparatory to planting.

3. *Formaldehyde.*—This chemical has long been used in liquid form, properly diluted, as a seed disinfectant for the smuts of cereals. The solution is prepared by adding 40 per cent formalin (formaldehyde) to water at the rate of 1 pint to 30 to 40 gallons of water. Seeds to be treated are placed in a cloth bag and immersed in the solution. Most vegetable seeds are treated for 10 minutes, after which they are rinsed in water or a milk of lime solution prepared by adding 1 pound of quicklime to 10 gallons of water.

4. *Other Chemicals.*—Red oxide of copper in powder or dust form has been shown to be an effective fungicide for damping-off. It has sticking qualities that enable it to adhere remarkably well to the seed coat. Small quantities of seed may be treated by shaking the seed in a closed container to which red oxide of copper has been added at the rate of 1 teaspoonful of dust to each pound of seed to be treated.

Copper carbonate may be used effectively for the control of the damping-off of vegetable and flower seedlings. It is also used as a treatment for wheat smut. It is used as a dust, 2 to 3 ounces being sufficient for a bushel of seed. Larger amounts are not injurious. Seed may be treated several months before planting.

Various organic mercury compounds are used effectively in treating diseases borne on the seed coat. Some are dusts; others liquids. Some of the more popular ones are Uspulun, Semesan, Ceresan, and Germisan. Doubtless there are many others that are equally as effective.

Diseases Borne within the Seed Coat.—Certain plant diseases are caused by organisms that may be borne within the seed coat. The fungus that causes blackleg of cabbage and related plants is of this class. Treatment of infected seed with chemicals is effective only to the extent of killing the parasite on the outside. Treatment with hot water at a temperature of 122°F. for 25 to 30 minutes is effective in destroying the parasite on the inside as well as on the outside. The treatment is severe, and only seeds with strong vitality withstand it. If seeds are known to be infected, they should be discarded. The treatment is recommended only as a precautionary measure.

The fungus that causes anthracnose of beans is also borne within the seed coat, and hence seed treatments used to control surface-borne parasites are not fully effective. The bean seed is killed by hot-water treatments of intensity and duration that kill the fungus, and means of control other than seed treatments must be relied upon.

Virus Diseases.—In addition to the diseases caused by visible parasites, on or within the seed coat, there are also virus diseases. Little is known about the real nature of these, and no good classification of them has been made. They may be transmitted from one plant to another in different ways. Most viruses are not carried by the seed from diseased plants. Certain mosaics of legumes, lettuce mosaic, and mosaic of wild cucumber are, however, transmitted by the seed from infected parent plants.

The cause of virus diseases is not known, and hence no treatments are recommended that will make infected seeds safe for planting.

SEED GROWING

Since the discoveries of sex in plants by Camerarius and of the fundamental laws of inheritance by Mendel, it has been recognized that considerable knowledge and skill are required for the intelligent production of seed. The inherent possibilities of mature plants are contained in the seeds from which they grow, and from this fact comes the old saying that nothing is so costly as cheap seed.

Methods of Production.—The commercial production of vegetable seed involves two distinct steps: first, the production and standardization of stock seed; and, second, the growing of these seed on a large scale for the trade.

The stock seed are usually grown by the seed firms themselves, on their own grounds. Methods employed are determined, to some extent, by the characteristics of the plant under considera-

FIG. 30.—Lettuce seed are harvested and stored in large tarpaulins until threshed.

tion. The same procedure would not be followed, for example, on the bean and the muskmelon. The breeding plots are kept under careful observation; all weak, diseased, or offtype plants are removed whenever noted; and special emphasis is placed on vigor, season, quality, uniformity, and trueness to type.

Stock seeds produced by the seedsman are then sent out to contract growers. These are generally farmers of a part of the country where seeds of the crop in question can be grown. Different sections are adapted to the growing of different kinds of seeds, and these may be remote from the centers of commercial production of the same vegetables or flowers. The field men for the seed firms visit the farms of the contract growers regularly and make inspections of the crops. All offtype plants are removed, a procedure known as *roguing*, and the purity and trueness to type of each variety is observed. In species that hybridize readily, each variety must be planted far enough from all other varieties to insure that it will not be crossed. The crop of seed is then harvested and sent to the seed dealer, who offers it for sale the following season and plants a plot on his trial grounds at the same time, as a check.

Classes of Seeds.—The botanical purity of some seeds can be determined by an examination of the specimens themselves; others can be judged only by the plants that they produce. Most vegetable and flower seeds fall within this latter class. The most careful examination reveals no consistent differences between the seeds of cauliflower and cabbage, for example. Marked resemblance exists between the seeds of carrot and parsley, cucumber and cantaloupe, onion and leek, and pepper and eggplant. Growers cannot determine the quality of a sample of seed by inspection. They must depend partly on the reliability of the seedsman from whom they are purchased, and they may, in addition, run seed tests to determine the value of a given sample.

Seed Testing.—The testing of seed involves two separate considerations:

1. *Mechanical Analysis.*—A sample of seed may show a considerable amount of inert material or even dead seeds of some other kind. It may contain viable seeds of other plants, including noxious weeds. The Russian thistle was first introduced into the United States mixed with a shipment of wheat seed; it has become one of the most dreaded plant pests. The purity of any sample of seeds from this viewpoint must be determined, especially when seeds are bought in bulk and in large quantities.

2. *Physiological Examination.*—There are three phases to this step in the testing of seeds. (*a*) Germination tests are conducted, in which various methods may be employed. Moist cloth, saucers, and blotting paper may be used, but it is preferable to plant the seeds under the more normal conditions of a sand or soil plot. Regardless of the method, it is important to ascertain the percentage of seeds that germinate and the time required. Such information will determine whether seeds are to be planted normally, planted thicker than usual, or discarded. (*b*) Another

phase of the physiological examination involves testing the vitality of the sample. Vigor of growth of the young seedlings is a very important character. The young plant is dependent for a time on food stored in the seed, and for this reason healthy mature seed will give the plant a better start. Seeds may be sown in plots of clean sand and tested for vitality at the same time the germination test is being made. The presence of disease in the seed may be detected if the seeds are sown in sterile sand. Diseased seeds are frequently responsible for losses that could have been avoided by proper testing. Seeds

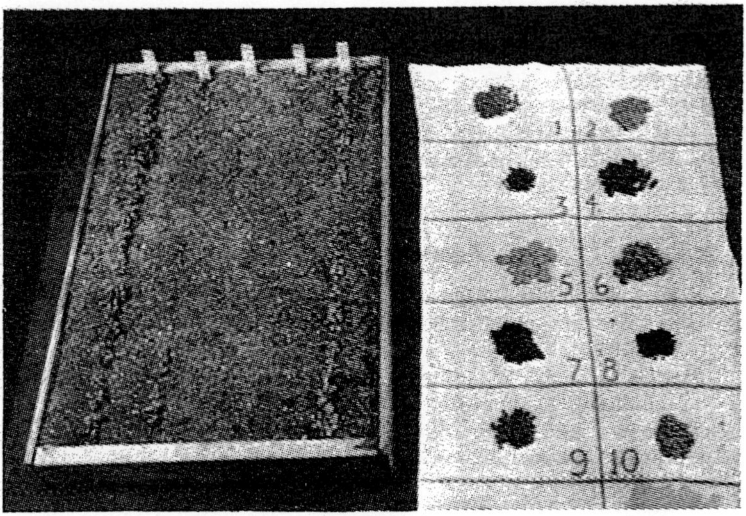

FIG. 31.—Two common ways of making a germination test—sand flat on left, rag-doll method on right.

with pronounced vitality will frequently yield a good stand under unfavorable conditions, whereas those that do not have such unusual vigor would scarcely germinate under the same conditions. (*c*) Additional information on the value of a given lot of seed for planting may be gained by growing a trial crop from a sample of the seed. This is done one season, and, if the resulting crop is satisfactory, seeds from the lot are used for planting during seasons that follow. The growing of such test crops may yield information relative to the presence or absence of seed-borne diseases, trueness to strain or variety, and the presence or absence of closely related species or noxious plants.

Questions

1. What are the various stages in the process of seed germination?

2. What is a seedling? What are the principal parts of the young seedling?

3. What are the differences between the root systems of seedlings of monocotyledonous plants and those of dicotyledonous plants?

4. What is the explanation for the phenomena whereby cotyledons of the bean are borne to the surface during germination, while those of the garden pea are not?

5. What conditions are essential for germination?

6. What is the relationship of vitality to viability? What factors influence the vitality of seeds?

7. To what extent do seeds vary in their storage requirements?

8. What are the causes of delayed germination of seeds? What can be done to overcome each cause?

9. What treatments are recommended for diseases borne on the seed coat? Within the seed coat? Virous diseases?

10. What information is obtained by a mechanical analysis of seed? By a physiological examination?

Suggested References

1. ADAMS, W. R.: The Influence of Soil Temperature on the Germination of and Development of White Pine Seedlings, *Vt. Agr. Exp. Sta. Bull.* 379, 1934.
2. BAKKE, A. L., H. W. RICHEY, and KENNETH REEVES: Germination and Storage of Apple Seeds, *Iowa Agr. Exp. Sta. Res. Bull.* 97, 1926.
3. BEATTIE, J. H., A. M. JACKSON, and R. E. CURRIN: Effect of Cold Storage and Age of Seed on Germination and Yield of Peanuts, *U. S. Dept. Agr., Circ.* 233, 1932.
4. BIOLETTI, F. T., and W. F. OGLESBY: Growing and Grafting Olive Seedlings, *Calif. Agr. Exp. Sta. Bull.* 268, 1916.
5. BORTHWICK, H. A.: Factors Influencing the Rate of Germination of the Seed of Asparagus Officinalis, *Calif. Agr. Exp. Sta. Tech. Paper* 18, 1925.
6. ——— and W. W. ROBBINS: Lettuce Seed and Its Germination, *Hilgardia*, **3**: 275–304, 1928.
7. BRENTZEL, W. E.: Seed Treatment, *N. D. Agr. Exp. Sta. Circ.* 56, 1935.
8. CLAYTON, E. E.: Increasing Stands from Vegetable Seeds by Seed Treatment, *N. Y. (Geneva) Agr. Exp. Sta. Bull.* 554, 1928.
9. COOK, HAROLD T., and JOHN A. CALLENBACH: Truck Crop Investigations, Spinach Seed Treatment, *Va. Truck Exp. Sta. Bull.* 87, 1935.
10. GOSS, W. L.: The Vitality of Buried Seeds, *Jour. Agr. Research*, **29**: 349–362, 1924.
11. JONES, LEON K.: Factors Influencing the Effectiveness of Organic Mercury Dusts in Pea-seed Treatment, *Jour. Agr. Research* **42**: 25–33, 1931.

12. KINMAN, F. C., and T. R. MCCLELLAND: Experiments on the Supposed Deterioration of Varieties of Vegetables in Porto Rico, with Suggestions for Seed Preservation, *Puerto Rico Agr. Exp. Sta. Bull.* 20, 1916.
13. MCKEE, ROLAND: Vernalization Experiments with Forage Crops, *U. S. Dept. Agr., Circ.* 377, 1935.
14. MUNN, M. T., and RACHEL E. MUNN: The Quality of Flower Seeds on Sale in New York, *N. Y. (Geneva) Agr. Exp. Sta. Bull.* 663, 1936.
15. ——— and ———: The Quality of Vegetable Seeds on Sale in New York in 1934 and 1935, *N. Y. (Geneva) Agr. Exp. Sta. Bull.* 664, 1936.
16. SPAETH, J. NELSON: A Physiological Study of Dormancy in Tilia Seed, *N. Y. (Cornell) Agr. Exp. Sta. Mem.* 169, 1934.
17. TUKEY, H. B.: Studies of Fruit Seed Storage and Germination, *N. Y. (Geneva) Agr. Exp. Sta. Bull.* 509, 1924.
18. ———: Seedling Fruit Stocks, *N. Y. (Geneva) Agr. Exp. Sta. Bull.* 569, 1929.

CHAPTER V

FORCING EQUIPMENT

Forcing is the growing of plants at a season in advance of that when they are normally grown. The common objective is the production of early fruits, vegetables, and flowers, for which there may be a special demand. The plants may be started from cuttings or from seed, and they may be grown to maturity or transplanted to the open field when conditions outside become favorable. Forcing structures include cold frames, hotbeds, greenhouses, and other similar equipment used in growing such plants.

TYPES OF FORCING STRUCTURES

There are several different kinds of forcing structures. The kind or species of plant to be grown, the length of time the structure is to be needed during a season, initial cost, operating expense, and other similar factors are considered in deciding upon which to use.

Forcing Hills and Plant Protectors.—In general, forcing hills are designed to protect growing plants in the field. Several types are used; the most elaborate one is made in the form of a box, usually 12 inches square and 12 inches high. The box is placed over the plant to be protected from cold or wind. A pane of glass is used for a cover. It serves to exclude cold, but at the same time permits the plant to be partially exposed to sunlight. On warm, sunshiny days the glass may be removed entirely. After danger of cold weather is past, the box is removed, but it may be held in readiness for use in the event of an unexpected late frost.

Other types of plant protectors are used for plants growing in the open. Small conical covers made of paper that is more or less translucent are satisfactory where it is desired to protect plants only during an emergency. These paper caps when placed over the plant are held in place by dirt that is piled up around the base. The same protection may be had on a limited scale by placing tin cans over the plants. Melon vines may be protected in a similar manner by placing small squares of glass over them, where the seeds have been planted and the plants caused to grow in a slight depression.

On a commercial scale, the use of equipment designed to offer protection in the field is restricted to plants that produce heavily of a valuable crop. The tomato and muskmelon are examples of such plants. It would be impracticable to use forcing hills in the growing of carrots or radishes; the unit return from an indi-

FIG. 32.—Conical paper covers used to protect tender plants in the field.

vidual plant of such crops is too small to justify the expense for labor and material.

Cold Frames.—Cold frames are designed primarily to protect plants from cold, but they are not heated artificially. They are used generally in the starting of vegetable crops and to a lesser extent in the handling of cuttings.

The standard frame is 6 feet wide and of any convenient length. It is commonly constructed of either wood or concrete, although other materials are sometimes used. Frames of wood are frequently made in such a way as to be movable; this provision makes it possible to set the frames up at different places each year and to store them during off seasons. A cold frame should be located on the south side of a building or other barrier wherever possible, because of the protection from north winds. It should extend in lengthwise direction from east to west. The north wall of the structure should be 6 inches higher than the south wall. This facilitates shedding of water when the frame is covered, exposes the interior more directly to the sunlight, and provides some protection from north winds. The floor of the

FIG. 33. —Cold frames used for hardening cabbage plants, prior to field planting.

cold frame should, of course, be practically level; otherwise it would be difficult to water the seedbed uniformly. The floor surface should be even with or slightly above the surrounding ground level in order to insure good drainage.

The most satisfactory cover for the cold frame is the standard sash. An individual sash is 3 feet wide and 6 feet long and has glass panes of appropriate dimensions embedded in its frame. Cold frames on which sash is to be used are constructed in units agreeing with these dimensions. The sash is placed lengthwise across the cold frame, the upper edge of which is provided with a shoulder to hold it in place. The standard sash is expensive; yet it is probably the most satisfactory cover. A frame covered with glass absorbs heat from the sun on clear days and retains it during the night and on cold days; it is possible in this way to provide temperatures that are more uniformly favorable for plant growth than would be the case if the frames were not so covered.

Various other materials are used as covers for cold frames. Screen wire imbedded in a transparent material similar to

FIG. 34.—Tomato plants growing in special beds covered with cloth, prior to being transplanted to the field.

cellophane makes a satisfactory cover. This material is usually tacked on frames of dimensions that permit of convenient handling. Different grades and weights of cloth that range from heavy duck to light domestic are also used. The untreated cloth may be used, but treating the material with hot linseed oil or melted paraffin increases its durability, makes it more nearly waterproof and airtight, and renders it more effective in protecting the frame during unfavorable weather.

Cold frames are used primarily in protecting plants against a few degrees of cold, usually in early spring. They are also useful in providing protection against wind and excessive rainfall, and in the hardening of plants prior to transplanting to the field, a practice that is discussed in the chapter on transplanting. In some places, certain crops are started in cold frames and, when the weather permits, the frames are removed and the crops receive field culture.

Hotbeds.—A hotbed is not greatly different in construction or dimensions from a cold frame. A protected location should be selected for it. Standard sash, treated and untreated cloth, and other materials are used for covers. The essential difference between the two is that the cold frame receives no heat except that from the sun, while the hotbed is provided with artificial heat. Plants may be grown in hotbeds at seasons when it would

FIG. 35.—Hotbeds heated by hot water. Heating pipes attached to walls of beds.

be too cold for them in cold frames, and the season of profitable use of such a structure is longer. Hotbeds are commonly heated during very early spring, but, as the weather warms up and plants need less protection, no heat is applied and the bed is used in much the same way as a cold frame. Heating of hotbeds is accomplished in four principal ways:

1. *Hot Water or Steam.*—Where hotbeds adjoin a greenhouse that is heated by steam or hot water, the heating pipes may be extended into the beds also. Other provisions are sometimes made for steam or hot water. The pipes are usually placed about 5 or 6 inches below the seedbed surface. Where it is desired to protect plants against an occasional late frost or freeze, and where it is not necessary to warm the soil, the pipes

may be suspended along the inside walls above the level of the seedbed. Steam- or hot-water-heated hotbeds are very satisfactory because the temperature can be regulated accurately.

2. *Manure.*—The heat given off in the decomposition of organic matter can be used as a source of heat for hotbeds. Animal manures are used more commonly than any other materials, and fresh manure from grain-fed horses is considered best. Cow manure is not satisfactory, largely because its texture does not permit of rapid decomposition, on which the generation of heat depends. It may be used, however, provided it is mixed

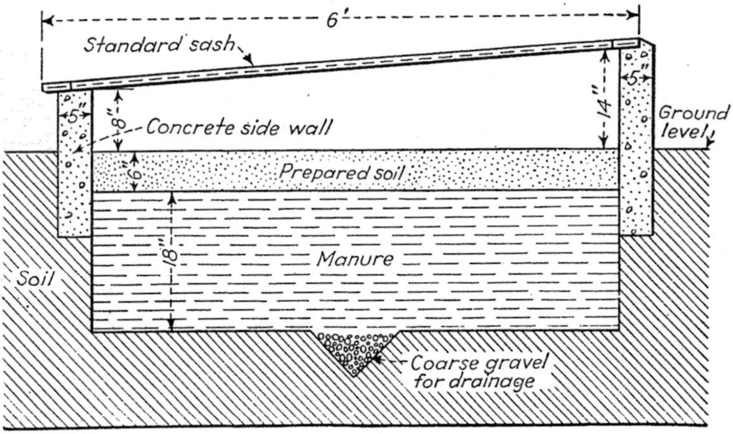

Fig. 36.—Cross-section drawing of a manure-heated hotbed.

with hay, straw, or some other coarse fibrous material that will favor aeration.

Practices vary in the construction of manure-heated beds. Generally the fresh manure is piled outside the hotbed. Water is added as it is being stacked, in order to encourage decomposition and to prevent "firefanging."[1] Within 4 or 5 days the temperature will run quite high, possibly 150 to 160°F., and will gradually recede. After it has reached the peak and dropped to 135 or 125°F. the manure is ready to be packed into the hotbed. Preparatory to placing the manure the entire area within the bed should be excavated to a depth of from 18 to 24 inches, depending on the length of the period during which it is anticipated heat will be needed. The manure should be put down in layers about 4

[1] Firefanging is caused by a fungus. Its presence is indicated by areas in which the manure is ashy gray or almost white in color.

inches thick and each layer packed well, especially along the edges and in the corners. When the required amount of manure has been placed in a pit, a layer of good soil 4 to 6 inches deep is spread smoothly over the top. This constitutes the seedbed. In no case should the seedbed surface be lower than the level of the surrounding ground. A good manure-heated hotbed in which the manure is 24 inches deep will exert a pronounced heating influence for 8 or 10 weeks. Those in which less manure is used will last a correspondingly shorter time. In no case is it advisable to use less than 12 inches of manure. The greatest heating effect is at the beginning of the period, and the temperature gradually subsides. Hence this type of hotbed is more satisfactory for use in the spring than in the fall.

Manure hotbeds are cheap and easily constructed. They require no great amount of care. The main objection to them is that there is no way to control the temperature accurately, once the bed is in operation. It may be lowered by ventilation but cannot be increased materially in case of unexpected cold weather.

3. *Flue Heat.*—In localities where fuel is cheap, so-called "flue-heated hotbeds" are used. A firebox is located at one end of the bed, and the heat is carried beneath the bed in one or more tile flues, which open into a low chimney at the other end. Two lines of tile properly spaced give a more uniform distribution of heat from one side to the other than if only one line is used. The hot gas and smoke in passing under the bed create the heating effect. Better draft can be created if the hotbed is located on a gentle incline, so that the firebox may be placed at a lower level than the chimney. Such a location may also facilitate drainage from the firebox during rainy weather. The length of the hotbed should be divided into a number of sections, depending on the slope, and the seedbed surface of each section should be level in order to facilitate uniform watering.

Cheap fuel is essential for the practical operation of flue-heated hotbeds. Wood, gas, and crude oil are some of the materials used.

4. *Electricity.*—It is expected that as electricity becomes more generally available it will be used to a greater extent in heating hotbeds. A special lead-covered heating cable is now available for the heating unit. It is laid back and forth across the bed at

intervals of 6 or 8 inches. Soil is added to make a seedbed 4 to 6 inches deep over it. A thermostat may be used to control the temperature at which the current will cut on and off. It may be regulated so as to provide a favorable temperature range for particular kinds of plants being grown. This eliminates some of the cost of labor in caring for a hotbed. On a very limited scale, light bulbs suspended in the air within the hotbed may be kept burning to keep the temperature above the danger point during short cold periods.

FIG. 37.—Flexible electric heating cable laid in hotbed, to be covered with soil. (*Courtesy, P. T. Montfort, Agricultural Engineering Department, College Station, Tex.*)

Greenhouses.—The glass-covered house represents an improvement over the cold frame and hotbed and is superior to the other forcing structures for starting plants. Temperature may be controlled more accurately, ventilation regulated more perfectly, and arrangements more convenient for work provided. Greenhouses are constructed in many sizes and types. The cost need not necessarily be exorbitant. They are sometimes made by providing low walls and using standard sash for the roof. In design there are three distinct types of greenhouses:

1. *Lean-to Greenhouses.*—This type is constructed on the side of a building. The roof slants in one direction. A lean-to

greenhouse should be located always on the south side of a building. Its length should extend in an east and west direction. When so located, the roof will necessarily slope toward the south. This is desirable, since it permits better exposure to the sunlight in the late winter and early spring.

2. *Three-quarter-span Greenhouses.*—In this type of greenhouse, three-fourths of the surface of the roof slopes in one direction, usually toward the south; one-fourth slopes in the opposite direction. The ridge of the roof is hence off-center. The southern slope of the longer span permits favorable exposure to sunlight; the quarter span makes it possible to ventilate more

FIG. 38.—View of a good type of greenhouse. Note curved eaves.

effectively. The three-quarter-span greenhouse should extend lengthwise from east to west.

3. *Even-span Greenhouses.*—The roof in this type slopes evenly in two directions, the roof ridge being above the center of the greenhouse. Possibly the best exposure to sunlight and heating effect is obtained if the lengthwise direction is north and south.

Hot air, hot water, and steam are used in heating greenhouses. Steam is quite effective, though expensive. Hot-water heating is probably the most desirable, especially in sections where continuous heat is required. Burning gas or other fuel in the greenhouse is sometimes practiced where heating is required only at intervals. It is objectionable because of the difficulty of heating uniformly and also because of the possibility of damage to plants by fumes. These considerations are serious where prolonged heating is necessary. A better way is to conduct the hot air through the greenhouse in tile flues.

Slatted Frames.—Frames that are constructed so as to provide partial shade for tender young plants are known as *slatted frames* or *semishades.* These differ from other forcing structures in that they are designed to protect plants from excessive heat or hot sun instead of cold.

A satisfactory slatted house can be made by spacing 2-inch strips horizontally at intervals of 2 inches over a framework of convenient height. One-half of the area is thus covered. It

FIG. 39.—Slatted frame constructed so as to be movable.

may be advisable to strip one side and the ends as well as the top; in this case, however, the strips should not extend nearer than within 1 foot of the ground, in order to allow proper ventilation. Frames that are 12 feet long, 6 feet wide, and 5 feet high, if properly braced, can be handled easily by two men. There are advantages in constructing them so that they can be moved. The seedbed may be located in a different place each succeeding year, and it may be prepared and planted in advance, after which the slatted frame is placed over it.

The Solar Propagating Frame.—A structure known as the *solar propagating frame,* that operates on the principle of receiving bottom heat from the sunlight, has been developed. It has been used successfully for rooting cuttings of citrus and other subtropical plants. The *box frame* is similar to the solar frame in principle but is different in construction. It has been used successfully in propagating blueberries. An essential feature of these structures is the elevated cutting bin. The frames may vary in size but are frequently constructed with the following approximate dimensions: height, 40 inches; length and width, 72 by 30 inches; depth of bin, 8 inches. The floor of the propagating bin constitutes a false floor for the rooting media, underneath which is an enclosed dead-air space. An apron is constructed on one of the long sides in such a way as to permit heat to be absorbed from sunlight by materials provided for the purpose. Water, rocks, and bricks are frequently used, being covered with a glass sash to hold the heat. The heat absorbed during the day and given off at night tends to keep the temperature of the rooting medium higher and more uniform than would otherwise be true. Screen wire properly supported is frequently used for the false floor in order to permit a ready absorption of heat from below.

USES OF FORCING STRUCTURES

The principal uses for forcing structures include vegetative propagation, starting of plants, and the growing of plants to maturity.

Forcing Structures and Vegetative Propagation.—Certain kinds of vegetative propagation work can be carried on most successfully in a forcing structure. Earlier sweet potato slips can be produced by bedding the sweet potatoes in a hotbed. Florists and nurserymen follow a practice of rooting cuttings in the hotbed, cold frame, or greenhouse. Conditions essential to the success of cuttings may be provided more easily in a forcing structure than in the open; and hence better results are obtained. Slatted frames are useful for rooting softwood and herbaceous cuttings during the summer months.

Successful bench grafting depends largely upon callusing, which in turn is influenced among other things by temperature. By virtue of the favorable temperature provided in a hotbed or greenhouse, bench grafting may be done throughout the winter.

Forcing Structures for Starting Plants.—Seeds of some species are invariably planted directly into the field where the plants that develop from them grow to maturity. Turnips, radishes, and beans are examples of such crops. Another class of crops may be started in either of two general ways: They may be planted directly into the field and allowed to grow to maturity without being moved; or the seed may be planted in a forcing structure and the resulting seedlings transplanted to the field after they have made sufficient growth. Tomato, cabbage, onion, pepper, and lettuce are examples of crops of this latter class. Practical considerations determine whether such crops are planted directly in the field or in a special seedbed and the seedlings transplanted

FIG. 40.—Interior of greenhouse used in propagation work. Plants in flats in foreground. Hardwood cuttings in bed in center.

to the field. There are several advantages, however, to be derived from starting plants in a forcing structure:

1. The grower is enabled to produce a marketable crop of a frost-tender species, as the tomato, much earlier than if the seed were planted directly in the field. Young seedlings 1 month or 6 weeks old are available for planting at a time when otherwise it would be necessary to plant seed. In any case, whether the crop is for home use or for market, the matter of earliness of maturity is most important.

2. The practice makes it possible to lengthen the favorable growing season of a locality and thus gives a tender species time to mature where the season might otherwise be too short. The

eggplant, for example, requires a long, hot growing season; it cannot be grown successfully in some sections of the North unless its season is lengthened artificially by starting the plants in a hotbed or greenhouse. Cabbage, on the contrary, requires a cool growing season. In the South, the weather often becomes too warm as it approaches maturity for the development of highest quality. This can be offset by starting the plants earlier and thus causing the cabbage to mature at an earlier date, when cool weather is expected. For the fall crop of cabbage in the South, the crop must be started early in order to mature before the winter freezes. Since the weather is too hot for optimum growth of the tender young plants in the open at that time, the slatted frame may be used to a good advantage.

3. The crop occupies the land for a shorter period of time and the grower has opportunity to make more effective use of a certain area. Two or more crops may be grown on the same land within a given time instead of perhaps one. A green-manure crop may be given time to make better growth, or possibly to decompose more completely, if the plants that are to follow do not have to go into the field so early.

4. Young plants can be given better care in a specially prepared seedbed; the weak ones may be culled out, and the resulting stand in the field will be more uniform both in size of plants and in spacing. Watering and insect control are more easily accomplished in a seedbed than in the open field. Conditions in a seedbed encourage the growth of good plants, since the detrimental effects of drought, excessive rainfall, or insect damage may be largely overcome.

5. The saving effected in seed may be considerable. Some vegetable and flower seeds are comparatively expensive, and it is desirable to make economical use of them. When seeds are planted in beds, every strong plant may be used; while if seeding is done directly in the field, the seed must be planted thickly and the seedlings thinned if a good stand is to be obtained.

It does not follow from these observations, however, that all seed should be planted in beds, and the seedlings transplanted. Some species may be transplanted readily, but still it is impracticable to transplant them on a commercial scale. Such is the case with turnips or carrots. The return from those crops does not warrant the expense involved in transplanting. There are other crops that do not stand transplanting well. Corn and watermelons are examples of such plants. These may be started in a forcing structure provided they are planted in pots and later shifted to the field without breaking the soil from about the roots.

FIG. 41.—View of extensive area of slatted frames, used in growing asparagus fern.

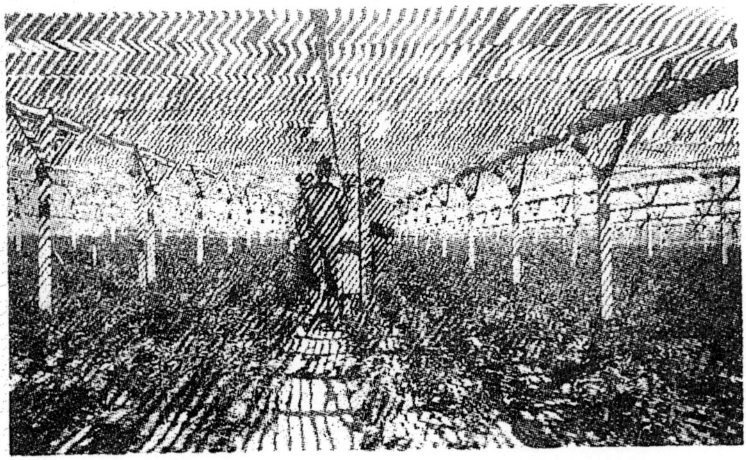

FIG. 42.—Interior view of area covered with slatted frames.

Growing Plants to Maturity.—Growers have found it practicable to grow some species of vegetables and flowers to maturity in forcing structures. The practice has given rise to the so-called "forcing industries." A great variety of cut flowers are obtainable from the florist practically every month of the year. Although some of these are shipped in from localities where they may be grown in the open, many of them are grown in greenhouses. Tomatoes and cucumbers are popular vegetable forcing crops. They are grown largely in regions where the winters are long and severe, and are planted so as to mature when crops cannot be grown outside. The greenhouse is the most important structure for growing forcing crops, although the hotbed and cold frame are used to a limited extent.

In some regions where winters are mild, plants, such as asparagus fern, are grown to maturity under slatted frames that protect against both heat and cold at different seasons.

Canned goods of high quality, vegetables preserved by quick-freezing, and fresh vegetables made available by good transportation facilities, the refrigerator car, and cold storage have all tended to discourage the forcing of vegetable crops. Many greenhouses that formerly grew lettuce or tomatoes now stand idle as a result of competition of lettuce and tomatoes that are grown as truck crops, shipped perhaps a thousand miles and sold at prices with which the greenhouse producer cannot compete. This type of competition tends also to discourage the forcing of cut flowers, though not to the same extent.

MANAGEMENT OF FORCING STRUCTURES

The use of good equipment does not in itself carry with it the assurance of success. Successful operation depends also upon good management. This applies equally to the hotbed, cold frame, greenhouse, and semishade. Management involves the primary problems of heating, watering, ventilation, and control of insect and disease pests. It involves also the use of good soil and the practice of certain rules of technique with respect to the actual details of operation.

Soil.—Sand is used generally in beds where cuttings are grown. Peat and rice ashes, alone or in mixtures with sand, are also used. For seedling plants that are to be transplanted, it is important that good soil be used in the seedbed. The soil may determine whether the plants are stocky or spindling, vigorous or stunted, normally developed or excessively luxuriant. Soil likewise influences directly the vigor of plants that are to grow to maturity in the forcing structure. It should be fairly fertile, of good physical texture, well aerated, and relatively free from insects and disease organisms.

1. *Soil Preparation.*—Specially prepared soil should be provided for seedbeds. The most valuable source is from compost beds. Compost is made by stacking alternate layers, about 4 inches thick, of good loam soil and barnyard manure until the pile is 4 or 5 feet high. It may be of any convenient width and

FIG. 43.—Electrical soil sterilizer.

length. The top should be concave so as to hold water that will soak into the layers and encourage decomposition. The compost pile is prepared from 1 to 2 years in advance of the time it is to be used. In the meantime, however, it should be spaded vertically and restacked so as to blend the loam and manure. This should be done every 3 or 4 months. A mixture of compost, sand, manure, and loam makes an ideal soil for use in seedbeds or for potting.

2. *Soil Sterilization.*—Insects and diseases in the seedbed soil may cause losses in several different ways. The seedling plants

may be killed while they are still in the bed, or they may become affected there, but with the trouble not showing up until the plants approach maturity in the field. The seedbed may be a source of infestation whereby insects and diseases are carried to the field on the roots and in adhering soil of the transplanted plants. Such pests once introduced may become established permanently and discourage the growth of crops in future years.

FIG. 44.—Soil-sterilizing unit, showing electrical heating elements.

Heat was first used in sterilizing soils, and it is still a popular treatment. A fire may be burned over the seedbed area, or steam may be admitted to the soil until all organic life is killed. Electricity is used effectively in sterilizing soils by a process known as *soil pasteurization.* The soil in a closed container is heated by electricity to a temperature of from 150 to 160°F., and the temperature maintained for several hours. Plants grow better in soil sterilized in this way than when the soil is overheated.

Chemicals may also be used in sterilizing soils. Formaldehyde dust is effective in controlling most fungous diseases when mixed at the rate of 8 ounces of 6 per cent dust to 1 bushel of soil. Nematodes are killed by using 16 ounces of dust for a bushel of soil. Liquid formaldehyde solution, prepared by adding 1 pint of 40 per cent formalin to 30 gallons of water, applied at the rate of ½ gallon per square foot, and dilute solutions of mercuric chloride (0.1 per cent) are also effective soil disinfectants. Zinc oxide spread over the seedbed surface prevents the spread of certain diseases. A red oxide of copper spray is effective in

FIG. 45.—Storage racks for pots.

controlling damping-off fungi above ground on seedlings, and if it soaks into the soil the disease is discouraged below the surface.

Very favorable results have been obtained recently in the use of chloropicrin and carbon bisulphide as soil fumigants for the control of root-knot nematode and other soil pests.

Watering.—Uniform growth of plants depends on a uniform supply of soil moisture. Much of this must be applied artificially to plants in a special seedbed. Water that is free from objectionable salts should be used. Rain water is preferable. The seedbed should be prepared so that the water will soak in uniformly and not run to one part of the bed. When plants are grown in pots, the soil should not completely fill the pot. Instead, each pot should have a slight basin at the top in order to facilitate watering. The size and depth of the basin should be uniform for all pots of a lot.

Temperature.—Different kinds of plants require different temperatures for optimum growth. A suitable temperature for cabbage would be too cool for tomatoes. Plants of either species, when grown at a temperature slightly above optimum, are likely to make an excessively luxuriant growth. If the tem-

FIG. 46.—Vat for washing pots.

perature is too low, they make limited growth and become dwarfed or stunted. Sudden changes in temperature should be avoided.

Lighting and Ventilation.—Normal growth of plants depends on the proper amount of light. In the location and construction of the seedbed it should be remembered that light is an important factor. Shade created by an adjacent building, trees, high side walls, or crowding of plants tends to cause succulent and spindling growth. Such plants are undesirable from every standpoint.

Ventilation is a means of admitting fresh air to plants. It is the most effective way of regulating conditions of temperature and humidity. A combination of high temperature and high humidity is objectionable, because it encourages disease, especially damping-off. Plants that grow in an atmosphere that is very humid are likely to make a slender growth and become excessively succulent and tender. The high water content of such plants is probably associated with the low transpiration rate under the conditions of high humidity.

Questions

1. Define forcing.

2. What are the principal forcing structures?

3. What are the main uses of the cold frame? The forcing hill?

4. What are the advantages of the hotbed over the cold frame? What covers are used for each? How are hotbeds heated?

5. What are the different types of greenhouses?

6. Tell how to construct a slatted frame. What are slatted frames used for?

7. Enumerate advantages to be derived from starting plants in a forcing structure for use in field planting.

8. What provisions are made for soil for forcing structures? How is compost prepared?

9. Why is it important to use only soil that is free from disease and insects?

10. What are the ways of sterilizing soil?

11. What precautions should be observed in watering plants? In regulating the temperature? In lighting and ventilation?

Suggested References

1. BEACHLEY, KENNETH G.: Combining Heat and Formaldehyde for Soil Treatment, *Pa. Agr. Exp. Sta. Bull.* 348, 1937.
2. FARRALL, A. W.: The Solar Heater, *Calif. Agr. Exp. Sta. Bull.* 469, 1929.
3. HORSFALL, JAMES G.: Pasteurizing Soil Electrically to Control Damping-Off, *N. Y. (Geneva) Agr. Exp. Sta. Bull.* 651, 1935.
4. ———: Combatting Damping-Off, *N. Y. (Geneva) Agr. Exp. Sta. Bull.* 683, 1938.
5. JOHNSON, STANLEY: The Propagation of the Highbush Blueberry, *Mich. Agr. Exp. Sta. Sp. Bull.* 202, 1930.
6. KADOW, K. J., and H. W. ANDERSON: Damping-Off Control, an Evaluation of Seed and Soil Treatments, *Ill. Agr. Exp. Sta. Bull.* 439, 1937.
7. KRONE, PAUL R.: The Reaction of Greenhouse Plants to Gas in the Atmosphere and Soil, *Mich. Agr. Exp. Sta. Sp. Bull.* 285, 1937.
8. NEWHALL, A. G., and M. W. NIXON: Disinfecting Soils by Electric Pasteurization, *N. Y. (Cornell) Agr. Exp. Sta. Bull.* 636, 1935.
9. PRICE, F. E., and C. J. HURD: Electric Hotbeds and Propagating Beds, *Ore. Agr. Exp. Sta. Bull.* 307, 1932.
10. STOUT, G. J., J. E. NICHOLAS, W. B. MACK, and D. C. SPRAGUE: Methods of Heating Hotbeds, *Pa. Agr. Exp. Sta. Bull.* 338, 1936.

11. SWINGLE, WALTER T., T. RALPH ROBINSON, and E. MAY, JR.: The Solar Propagating Frame for Rooting Citrus and Other Subtropical Plants, *U. S. Dept. Agr., Dept. Circ.* 310, 1924.
12. VINCENT, CHESTER L., and HARRY L. GARVER: Hotbed Construction, Electric and Manure Types, *Wash. Agr. Ext. Serv. Bull.* 203, 1935.
13. WYMAN, DONALD, and MAURICE W. NIXON: Electric Hotbeds for Propagating Woody Cuttings, *N. Y. (Cornell) Agr. Exp. Sta. Bull.* 618, 1934.

CHAPTER VI

ASEXUAL PROPAGATION

Asexual reproduction of plants differs from sexual reproduction in that vegetative plant parts, such as stems, roots, bulbs, and leaves, are used rather than seeds. In order to propagate a new plant from any of these, it is necessary that the vegetative portion either (1) produce new roots that will support it, as in cuttage and layerage, or (2) unite by callusing with another plant that supplies the root system, as in budding and grafting.

Types of Asexual Propagation.—The methods by which plants may be reproduced asexually are bulb propagation, layerage, cuttage, and graftage. Each of these types of asexual reproduction will be considered separately in subsequent chapters. The particular method to be followed should be determined by the adaptation of the plant to be reproduced. It should be borne in mind that plants propagated by one method do not differ in any basic respect from those that may be propagated by another asexual method. The simplest method that will give satisfactory results is always used in commercial work. For example, it would seem that if all plants could be propagated by cuttage, that method would be universally used in preference to other asexual methods. Cuttings of many plants do not root well, however, and there is the further consideration of not wanting some plants on their own roots. Different rootstocks are used in certain specific cases, because of their adaptation to soil conditions, or their resistance to the attacks of diseases or insects. The same plant may be grown on several different rootstocks, to meet certain local conditions.

Perpetuation of Varieties.—Asexual propagation is used generally in the propagation of fruit trees, small fruits, ornamental plants, and certain vegetable crops. Standard varieties of fruits have certain combinations of qualities of fruit and tree that make them desirable. The value of many ornamental plants is associated with unusual habits of growth or peculiar characteristics of the leaves or flowers, such as an unusual type of coloration or an especially attractive pattern of variegation. Many such varieties are self-sterile, and cross-fertilization is necessary for the production of seeds. Even varieties that are self-fertile may be freely cross-pollinated by bees, wind, and other agencies. The result is that the seed bears heritable

FIG. 47.—The Ness berry is a hybrid of the raspberry and the dewberry, and is propagated by asexual means. (*Courtesy, Texas Agr. Exp. Sta.*)

qualities from the plant that supplied the pollen as well as the one that matured the seed; consequently these plants will not reproduce the variety. On the other hand, parts of the parent plant itself, such as cuttings, buds, grafts, and bulbs, when caused to grow will reproduce the variety.

Standard grades and brands of fruit, so essential in efficient, orderly marketing, thus owe their existence largely to asexual propagation of varieties. Budding is the method that has been followed largely in the production of thousands of Elberta peaches, for example, all of which bear peaches that are essentially alike. The marketing of the fruit from a large number of trees of one variety is relatively simple when compared with the problem of marketing the fruit from seedling trees, no two of

FIG. 48.—The acorn shown at *a* is a hybrid of the live oak shown at *b* and the overcup oak shown at *c*. (*Courtesy, Texas Agr. Exp. Sta.*)

which produce fruit that is alike. Market grades are based largely upon a uniform product. Since fruits from individual seedling trees vary widely in ripening dates, size, color, flavor, and other qualities, the important relationship between asexual propagation of plants and marketing is apparent.

Certain horticultural varieties of fruits and vegetables do not ordinarily produce viable seeds; consequently vegetative parts are used entirely in their propagation. The banana, Washington Navel orange, common fig, and Irish and sweet potato are examples of plants of this class.

Production of Uniform Rootstocks.—Aside from the importance attached to asexual reproduction as a means whereby varieties may be reproduced and preserved indefinitely, it is of considerable importance in that it makes possible the production of uniform rootstocks for budding and grafting.

In the propagation of citrus and other plants a number of hybrids are used to a limited extent as rootstocks. Since most hybrid plants are sterile, asexual methods are of especial importance in their propagation.

Fig. 49.—Showing variations in seedling pecans. The Texas Prolific (*a*) and Onliwon (*b*) are seedlings of the San Saba variety (*c*); they can be perpetuated only by asexual means.

Difference in production from trees of the same variety in an orchard indicates that, aside from variations in soil, there may also be marked differences caused by the variable influence, direct and indirect, exerted by the rootstock on the scion. In research work with fruit varieties, it is particularly important that stocks used be uniform in order to avoid differences that may influence results and render them less significant.

It is inconsistent to exercise care in the selection of varieties to be propagated and to give no thought to the rootstock upon which they are to grow and through which they must obtain the plant-food elements for vegetative growth and production of fruit. Seedlings vary in habits of growth and fruiting and also in the qualities that determine their value as rootstocks, such as type of root system and resistance to drought, cold, alkali, parasites, or other external influences. Plants of standard varieties of most fruits are propagated by asexual methods in order to insure uniformity of the top; the performance of those that are grafted is influenced also by the rootstocks used, and the greatest uniformity of those can be assured only by propagating them asexually.

Bud Selection.—In connection with asexual propagation, the term *bud selection* is used to indicate the choice of outstanding parent plants from which to obtain propagation materials, such as scions, cuttings, tubers, and bulbs. Opinions vary as to the value of this practice, but a few facts are self-evident.

In the first place, there can never be any detrimental effect from using, as a source of propagating material, a good tree that produces abundant crops of fruit of high quality. The same statement would apply to plants of outstanding merit in other respects. The practice has much to commend it from the standpoint of insurance. But intelligent bud selection necessitates the keeping of tree-performance records; this additional expense and the fact that the progeny of a certain tree are not always so good as the parent are the chief drawbacks of the method. Trees produced from such buds may cost 5 to 10 cents more than others, which will add several dollars per acre to the initial cost of the orchard. The failure of the method to perpetuate certain types of plants and fruits has been due in most cases to the inability of propagators to distinguish between fluctuating variations and mutations or "bud sports."

1. *Fluctuating Variations.*—Within an orchard the trees of a single variety may produce varying amounts of fruit, and the fruit in turn may vary to some extent in size, color, time of maturity, or other qualities. In numerous cases, investigators have used such trees as parents in order to determine whether these differences might be perpetuated. High-producing and low-producing apple trees have not transmitted these qualities to their respective progeny, and similar results have been obtained with other deciduous fruits. It is generally conceded that most of these differences in performance are due to environmental factors, especially soil fertility and soil moisture. A tree that produces heavy crops of good fruit by reason of its favorable location cannot be expected to transmit these qualities to trees that are grown under more rigorous conditions.

2. *Mutations or "Bud Sports."*—It has been found that an individual bud on a tree or plant may give rise to some type of variation that is definitely fixed and can be transmitted by asexual propagation. This character or behavior has been considered to occur without definite rhythm or regularity, but it is much more common in some species than in others. The Washington Navel orange, which was known as the Bahia before its introduction into the United States from Brazil in 1870, is generally conceded to have originated as a bud sport. The Starking variety of apples came into existence as a result of asexual propagation of a sporting branch of an apple tree of the Delicious variety. Starking apples develop color at an earlier stage of maturity than the parent variety. Attempts to produce mutations artificially are still in the experimental stage.

The reason that mutations have been the source of much confusion is that they may be observed in several different stages of growth. Where such a variation shows up on a twig or branch, which may produce fruit of a deeper color or with some other noticeable character, it is not difficult to detect it and propagate it. In some rare cases, however, mutant buds have been used in propagation work and a whole tree produced that was different from the type of the variety and capable of transmitting this difference. There would be no positive method of determining this difference, however, except by propagation and field trial.

Value of Bud Selection.—It should be recognized that bud selection is not a means of improving a variety. It is regarded rather as a means of isolating superior strains of varieties and preserving them true to type. It simply involves precautions against the use of inferior sports or the confusion of similar varieties. Mutations occur that are not easily observed, and these, when propagated unknowingly, give rise to varieties that may closely resemble the standard parent and yet be inferior to it in one or more respects.

Selection may also be a means of avoiding the use of diseased propagating material. Several diseases are transmitted by the vegetative parts used in propagation. Some of these are easily detected at one season but perhaps not at another. Certification of sweet and Irish potato tubers for planting is based largely upon freedom from disease and trueness to variety. Large Irish potatoes may be no more valuable than small ones for planting purposes, except that size may indicate freedom from disease.

Questions

1. Why is asexual propagation also called "vegetative propagation"?

2. Why are fruits generally propagated by vegetative means?

3. What are the causes of variable production of trees of the same variety?

4. What is a sport? In what ways may sports differ from the parent plant? How can a variation in a part of a plant be definitely identified as a sport?

5. What is the chief value of bud selection?

Suggested References

1. Cummimgs, M. B., E. W. Jenkins, and R. G. Dunning: Rootstock Effects with Cherries, *Vt. Agr. Exp. Sta. Bull.* 352, 1933.
2. Drain, Brooks D.: Field Studies of Bud Sports in Michigan Tree Fruits, *Mich. Agr. Exp. Sta. Tech. Bull.* 130, 1932.
3. Shamel, A. D., C. S. Pomeroy, and R. E. Caryl: Bud Selection in Eureka and Lisbon Lemons and Progeny Tests of Bud Variations, *U. S. Dept. Agr., Tech. Bull.* 531, 1936.
4. ——— and ———: Deciduous-Fruit Improvement through Tree-Performance Records, *U. S. Dept. Agr., Farmers' Bull.* 1696, 1933.
5. Tukey, H. B.: Seedling Fruit Stocks, *N. Y. (Geneva) Agr. Exp. Sta. Bull.* 569, 1929.
6. ——— and Brase, K. D.: Random Notes on Fruit Tree Rootstocks and Plant Propagation, I, *N. Y. (Geneva) Agr. Exp. Sta. Bull.* 649, 1934.
7. ——— and ———: Random Notes on Fruit Tree Rootstocks and Plant Propagation, II, *N. Y. (Geneva) Agr. Exp. Sta. Bull.* 657, 1935.
8. ——— and ———: Random Notes on Fruit Tree Rootstocks and Plant Propagation, III, *N. Y. (Geneva) Agr. Exp. Sta. Bull.* 682, 1938.
9. Webber, H. J.: Variations in Citrus Seedlings and Their Relation to Rootstock Selection, *Hilgardia* **7**: 1–79, 1932.

CHAPTER VII

BULBS AND RELATED STRUCTURES

Commercial production of bulbs is a comparatively new industry in the United States. Bulbs of all kinds have been produced successfully for many years, but only within the last few years has the scale of production become important. Formerly most of the planting stock was grown in foreign countries, notably the Netherlands. In recent years the government has placed definite

FIG. 50.—A naturalized planting of narcissus bulbs.

restrictions on importations of certain foreign-grown stock in order to lessen the danger of introducing new diseases and insect pests. These restrictions, although they have been modified from time to time, have had the effect of encouraging domestic production.

It is difficult to determine how many bulbs are now grown for propagation purposes. Importations alone during one of the peak years amounted to 327 million bulbs. Since that time commercial producing areas have expanded tremendously in Florida, Texas, California, Michigan, Virginia, and other states, where a considerable quantity of the domestic supply is now produced.

Bulbs are used in two principal ways, for forcing and for naturalized plantings. Bulbs are the most popular plants for forcing in the home. They may be grown in bowls of water, in sand, or in soil. Commercial forcing is a very specialized business, requiring skillful management and accurate knowledge of the plant material to produce flowers at the time when they are wanted. Keeping the greenhouse a bit too cool or too warm will mean the production of flowers at a time when perhaps there is no demand for them. Bulbs are likewise popular flowers for naturalized beds. They are hardy and have relatively few common insect pests or diseases; they bloom at a time of the year when flowers are scarce, and they grow several years with a minimum of care.

Classification.—Herbaceous plants that persist through the winter by means of subterranean buds are sometimes known as *geophytes*. The buds may be borne on specialized stems, such as rhizomes, tubers, corms, or bulbs; or they may be borne on specialized root structures of various kinds. The primary function of most stems is to expose the leaves to light, and for this reason the metamorphosed stems mentioned above have been frequently mistaken for roots.

The buds that are borne on these subterranean organs serve to continue the growth of the plant in the succeeding season after having carried it through a period unfavorable for growth. The various subterranean parts, which may be leaves, stems, or roots, consist largely of storage tissue and contain quantities of reserve food material. In some cases such storage organs disintegrate after the reserve food has been utilized in spring growth, but in others the storage parts persist and increase in size from year to year.

The various subterranean structures may conveniently be considered in the following classifications:

A. Bulbs.
  1. Tunicate or layered.
  2. Scaly.

B. Corms or "solid bulbs."

C. Rootstocks.
  1. Rhizomes.
  2. Pips.

D. Tubers.
  1. Stem.
  2. Root.

Bulbs.—Modified stems, in which the central axis is vertical and much shortened and the fleshy leaf scales or bases are closely

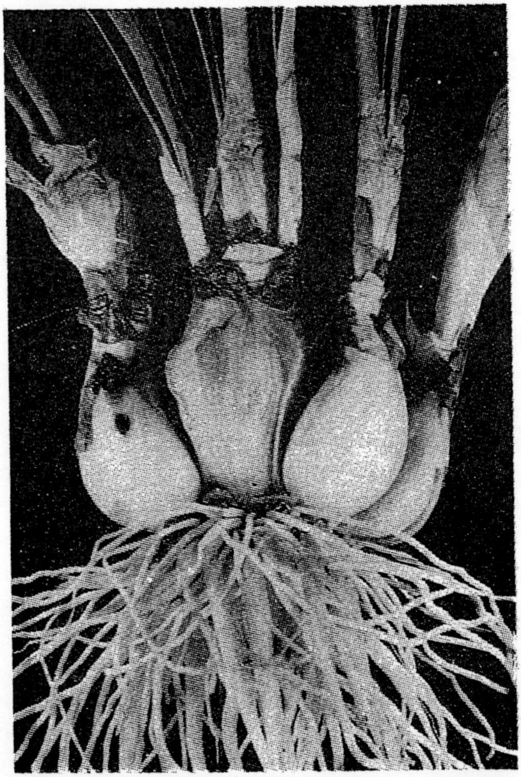

FIG. 51.—Cluster of narcissus bulbs, showing mother bulb and two slabs on the right and one on the left.

appressed, are called *bulbs*. In some bulbs the scales are continuous around the axis, forming a series of layers. In cross section these layers appear as concentric rings, as may be observed in the onion. Bulbs of this type are known as *layered* or *tunicate*. *Narcissus*, hyacinth, tulip, the bulbous iris, and onion are examples.

In other bulbs, the scales are not continuous but are rather narrow and fleshy and may be removed singly from the outer edges of the bulb. This type of bulb is known as *scaly;* the lilies are the outstanding members of this group.

In the bulbs, reproduction of roots takes place from the heavy basal stem plate. New shoots arise from the central axis, and

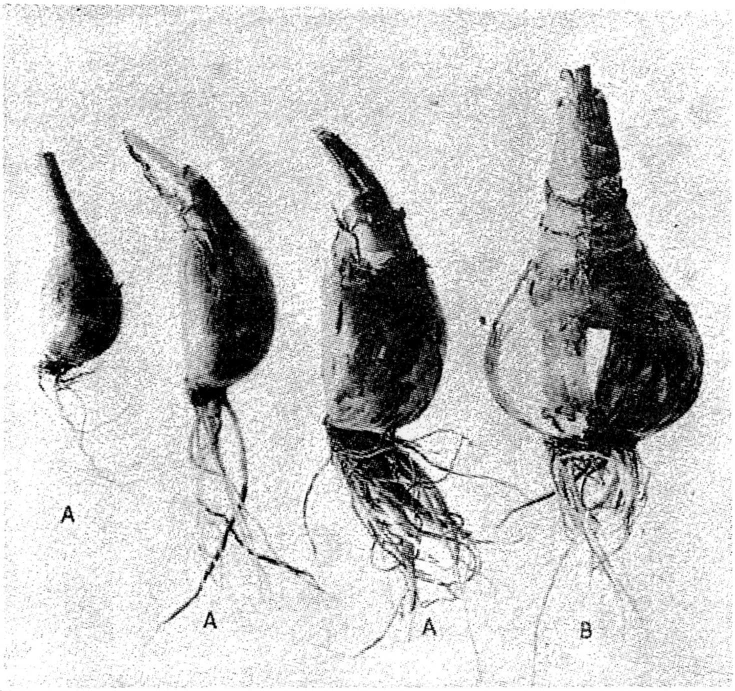

FIG. 52.—Golden spur daffodil bulbs: *A*, slabs; *B*, round. (*Courtesy, Bureau of Plant Industry, U.S.D.A.*)

new bulbs are produced from buds in the axils of the layers or scales. In some of the layered bulbs, like tulip, the old structure is consumed in growth of the plant, and a new bulb is reproduced each succeeding year. In other bulbs, such as narcissus, new bulbs are produced laterally but the old bulb continues to grow even after it has produced flowering shoots. The commercial reproduction of some of the more important bulbs will be considered in detail.

1. *Layered or Tunicate Bulbs.*—*Narcissus* comprises a group of very useful and popular flowering bulbs, including the types commonly known as *daffodils* and *jonquils*. These bulbs are probably the most satisfactory for naturalized plantings and are also used extensively for house flowers and for forcing in greenhouses. It has been estimated that in recent years the annual imports of daffodils amounted to 40 million bulbs, with an equal number of Paperwhites and other varieties. The value of these

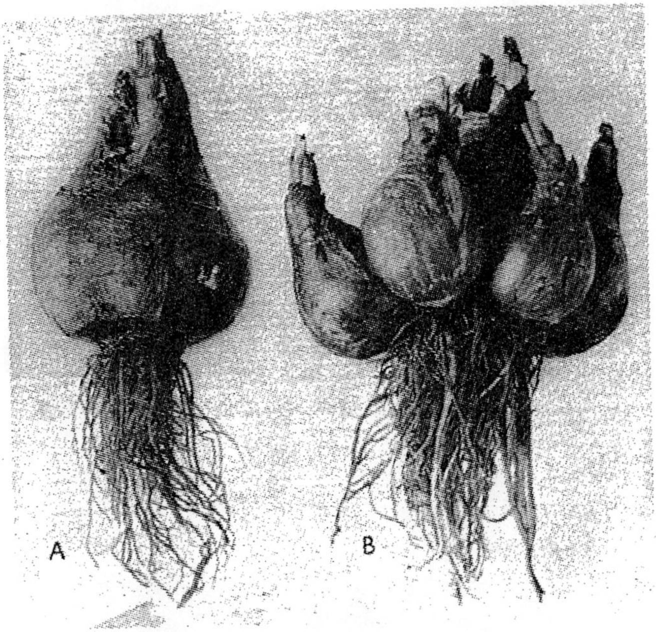

FIG. 53.—Daffodil bulbs: *A*, double-nosed type; *B*, mother bulb and slabs. (*Courtesy, Bureau of Plant Industry, U.S.D.A.*)

bulbs was about $2,500,000 and that of the cut flowers produced was more than twice that amount.

The normal cycle of reproduction of the narcissus requires a period of 3 years. The mother bulbs, as they reach maximum size, develop buds in the axils of the layers. These buds, still attached to the basal plate, continue to develop, forming daughter bulbs or splits, which may be separated easily from the mother bulb at the end of the growing season. These splits, separated and replanted, become in successive years, round, double-nosed, and mother bulbs, to complete the cycle of development. As the mother bulb is used up in growth of the new splits it is renewed from the central axis by the production of three or more layers and a new flowering shoot each year. The Paperwhite and the Chinese Sacred lily (types of narcissus) behave in similar fashion, except that a greater number of splits are produced and at an earlier stage of development of the bulb.

The various types of narcissus may be propagated also from seed. The time required for seedlings to reach the stage of maturity when splits are produced is usually 7 years.

Hyacinth propagation has been considered more difficult than that of most other bulbs, but leading authorities state that it is accomplished with reasonable care. Hyacinth bulbs are propagated in a fashion similar to that used with the narcissus, except that the old bulb itself splits up into several rather large ones, and a ring of small, flat bulbs forms around the basal plate. The chief disadvantages of the natural method of propagation are the lack of uniformity in the planting stock and the difference in time of maturity of the bulbs.

Artificial propagation is recommended for these reasons and also because a much greater number of bulbs can be produced from the original mother bulbs. There are several methods of artificial treatment, all of which involve the cutting of the mother bulb to cause an increased number of new bulbs to form. The three common methods are *scooping*, *scoring*, and *coring*. In the first method a concave cut is made to remove the entire basal plate and to cut through the basal portion of each layer of the bulb. In scoring the bulb, three cuts are made across the base, each cut crossing through the center of the bulb and each cut equidistant from the others. The cuts are made perpendicular to the basal plate and deep enough to extend through the growing point. In the coring process a core is removed from the central portion of the bulb, in the vertical axis, so that the entire growing point is removed.

The scored or cored bulbs are placed in dry sand or soil within a greenhouse or in the open field. Moisture will cause molds and eventual rotting of the bulbs, especially during the first 10 days. The scooped bulbs, however, are treated with a drying or disinfectant agent and allowed to callus in open air with little or no light. In either case after the bulbs have been callused, which will require about 2 weeks, they are placed on trays with wire bottoms. The bulbs should not touch each other in the trays, and every provision should be made for free circulation of air. Artificial incubation at a temperature ranging from 70 to 90°F. is necessary for the production and growth of young bulblets. The higher temperature should be maintained during the latter part of this growth period. Humidity in the incubation chamber

FIG. 54.—A clump of tulip bulbs which has developed from one mother bulb. (*Courtesy, Bureau of Plant Industry, U.S.D.A.*)

should be low at the start but increased later on to prevent excessive drying; toward the close of the process the air must be maintained near the point of saturation.

The period of incubation is approximately 2½ to 3 months. When roots begin to form, in the fall, the bulbs are set in the field. The plants are allowed to grow the following season until the mother bulb has disintegrated and the young bulbs will separate easily. These young bulbs are replanted the next season. Some few of them will reach maturity in 2 years, but normally 4 years will be necessary. The most desirable bulbs are those of largest size, before they begin to split; such bulbs will continue to blossom for several years.

The *tulip* is also a layered bulb, differing from the narcissus and the hyacinth in the fact that it has a small basal plate and a continuous brown coat. The bulb also has a characteristic shape. In the growth of the tulip bulb the basal plate gives rise to one large bulblet and several smaller ones. The old bulb disappears completely, and when the clump is dug there is one large bulb and one to four small flat ones. These smaller bulbs will require 1 or 2 years of additional growth before they in turn produce bulbs

FIG. 55.—A large marketable tulip bulb, on the right, and three splits, on the left, which will produce bulbs of flowering size the next year. (*Courtesy, Bureau of Plant Industry, U.S.D.A.*)

that will bloom. Bulbs of the three types, in order of their stage of development, are known as *flat bulbs* or *slabs*, *round* or *long-necked*, and *mature bulbs*.

The mother bulbs are planted in solid beds or in row furrows. The beds are usually 3 feet wide, with walks of 12 to 18 inches in width. The soil is removed to a depth of 4 inches from the entire bed, the bulbs are set in rows across the bed, and the soil is replaced. In the row method, furrows are opened 15 inches apart and 6 to 7 inches deep. The bulbs are set in each furrow as it is opened and are covered by the soil from the next furrow. Careful attention must be given to keeping the rows straight, since the bulbs are plowed out at the time of digging.

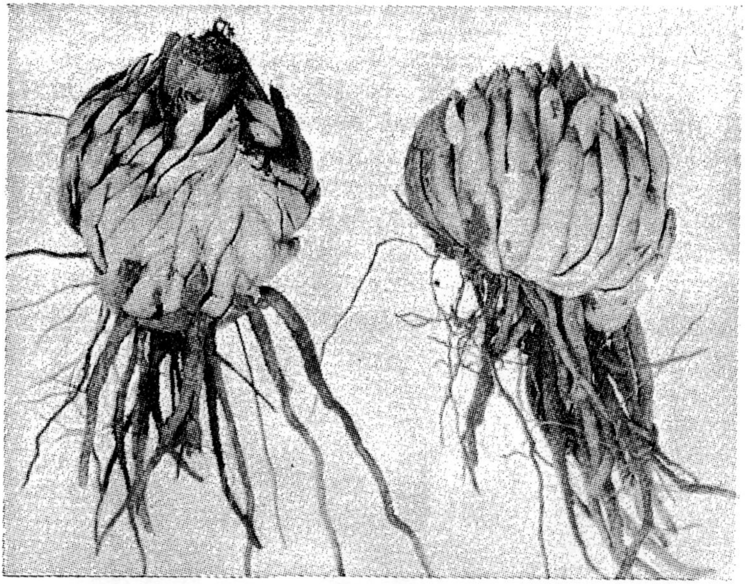

FIG. 56.—Typical lily bulbs. Old flower stalk is shown in bulb on left. (*Courtesy, Bureau of Plant Industry, U.S.D.A.*)

FIG. 57.—Lily bulbs: 1, mother bulb of flowering size; 2, longitudinal section showing old flowering stalk and new bulb.

When the bulbs are dug they are picked up in clumps, placed in field boxes, and carried into the house. They should not be left exposed to the sun and are usually covered with burlap in the field boxes. They are spread out to a depth of 4 inches in flat trays in the bulb house and dried for approximately 2 weeks. When the bulbs have dried they are cleaned by hand. The large

FIG. 58.—Lilies grown from scales. Planted in July and photographed the following April. (*Courtesy, Bureau of Plant Industry, U.S.D.A.*)

bulbs are removed from each clump, packed carefully in trays or boxes, and returned to the storage shelves. The small bulbs are piled up and cleaned later in a fanning mill. The bulbs should be handled carefully at all times to prevent bruising.

2. *Scaly Bulbs.*—The scaly bulb consists of a basal structure, frequently a rhizome, from which arise layers of scales that overlap to form the main part of the bulb. The young bulbs increase in size by the formation of new leaves from the growing point in

the center. The basal portions of these leaves flatten out to form new scales, and the process continues until the bulb is large enough to produce a stem.

Asexual propagation of *lilies,* the most important of the scaly bulbs, is accomplished by many different methods, for the reason that so many species are involved. The principal means of propagation include the use of scales, stems, leaves, aerial bulbs, and division of the old bulbs.

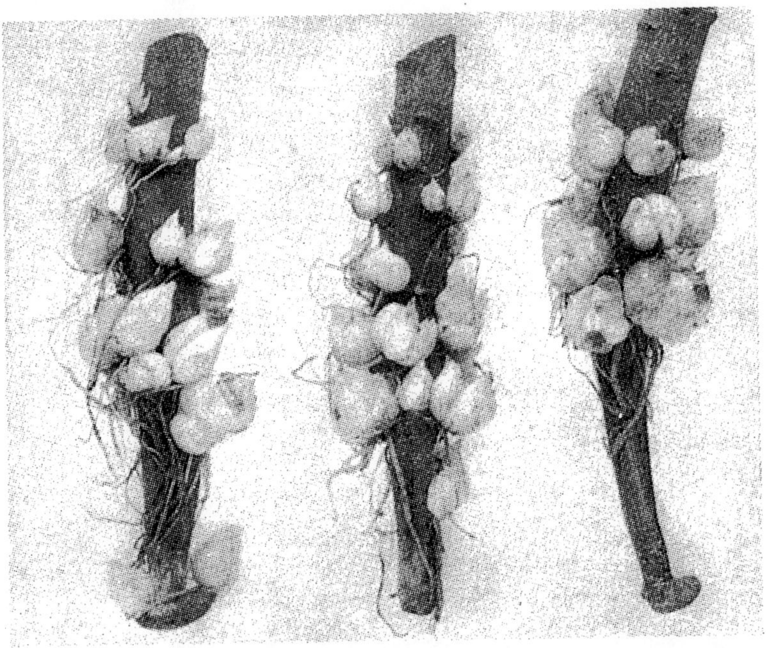

FIG. 59.—New bulbs have developed on the base of the lily stems which were heeled in the field from July 15 to October 15. (*Courtesy, Bureau of Plant Industry, U.S.D.A.*)

Scales are used successfully in propagation of many lilies. They must be removed from the bulbs in the late flowering stage for satisfactory results. The bulbs are dug at this period, the scales removed, and the bulbs replanted within a very few days. The number of scales that may be removed will vary considerably, but as many as 75 have been removed from one bulb without damaging it. The scales are planted in 3-foot solid beds, covered only 1½ to 2 inches, and by spring they will have produced bulblets on the concave sides. The normal number of bulblets is two on each scale, but an average of 1½, or 150 per cent increase, is considered very satisfactory. They may be left in the bed for a second season, by which time bulblets of considerable size will have been produced.

Stems of the various species may be caused to produce large numbers of new bulblets. The stems, with flowers usually cut off, are pulled from the ground and heeled-in. This should be done near the close of the blooming period. In a period of 35 to 40 days the stems are removed and will bear from 12 to 50 bulb-

FIG. 60.—The bulbs in this cluster have been allowed to grow undisturbed for several years. Note that the bulbs are crowded, and also note the large number of flowering stalks, which have been cut off. (*Courtesy, Bureau of Plant Industry, U.S.D.A.*)

lets, the number depending on the species. These may be removed and planted singly, or the entire stem may be set horizontally in a 3-foot bed.

Cuttings of the stems may also be made, with three or more *leaves*, or individual leaf cuttings may be made with heels of the stem. In either case, bulblets are formed in the axils and are then handled as described above.

Aerial bulbils are formed by several species of lilies. They occur in the axils of the upper leaves and may be removed soon after the flowering period. The bulbils may be set in beds and allowed to grow for 2 years, by which time about half of them will be producing flowers.

Division of the bulb occurs under natural conditions, and small increases may be obtained by digging the bulbs at intervals of 4 to 5 years for division. In commercial propagation, one of the other methods described will give more satisfactory results.

Seed are produced by almost all the lilies, and this method has the advantage of producing immense numbers of new plants. It also makes possible the continued production of new types by means of breeding. One of the serious handicaps in growing lilies from seed is the fact that some of the species produce seed that do not germinate the first season. In most cases, the seed are started in the greenhouse in the fall; the seedlings develop during the winter and are set in the field in the spring. Recently, considerable success has been obtained from field plantings in the Puget Sound section, where seed have been planted any time except from May to August with good results.

Corms.—A structure considered most conveniently with the bulbs is the corm or "solid bulb." This is not a true bulb but is designated by some writers as a shortened, thick, vertical rhizome. In cross or vertical sections the structure appears as a mass of solid, undifferentiated parenchyma. Roots are produced from the lower surface, as in the true bulbs, but there is no definite basal plate. Buds are formed from the upper surface and in turn produce small corms for reproduction. The outer covering of the corm is made up of the dried leaf bases. Corms are produced by gladiolus, crocus, and at least one species of iris.

The "bulbs" of *gladiolus* are usually placed in six commercial sizes. Two-year corms of the two larger sizes are best for ordinary planting, although some of the smaller ones will bloom. The larger corms, in addition to blooming, will be replaced by new bulbs, formed on top of the old ones, and these will blossom the following year. There will also be formed on the base of the new corm a large number of small cormels, ranging in size from a No. 6 down to the size of a pea, and in number from 12 to 100. The small ones are used for propagation of the variety.

The small cormels are replanted as soon as the corms are dug, or they may be dried and stored until the next year. The young plants will produce only foliage the next season, but when the corms are dug they will be about 1 inch in diameter and will have some new cormels on them. In the South, where the tops of the plants die in late spring, the bulbs are dug and stored through the summer and replanted in early fall. It requires two growing seasons to produce bulbs that are marketable.

The 1-year corms will form some flower shoots, but these should be removed as soon as detected, with every precaution to remove none of the foliage. In the succeeding years the bulbs are allowed to mature the blossoms, which are cut and shipped or sold locally.

Fig. 61.—A planting of gladiolus being grown for the production of new corms.

Producers who are located near large cities may find a demand for the flowers, so that two sources of revenue are found in the same crop.

Rootstocks.—A number of plants produce underground stems that show varying degrees of modification and are commonly used in propagation of these plants.

1. *Rhizome.*—This is a type of rootstock that is the least modified of the subterranean structures. A rhizome is simply a stem growing in a horizontal direction, slightly below the surface of the soil. It has the same general structure as all stems, including nodes and internodes, small scaly leaves, and axillary buds. Some plants of this type produce long underground parts and spread very rapidly. Johnson grass (*Holcus halepensis*) is a

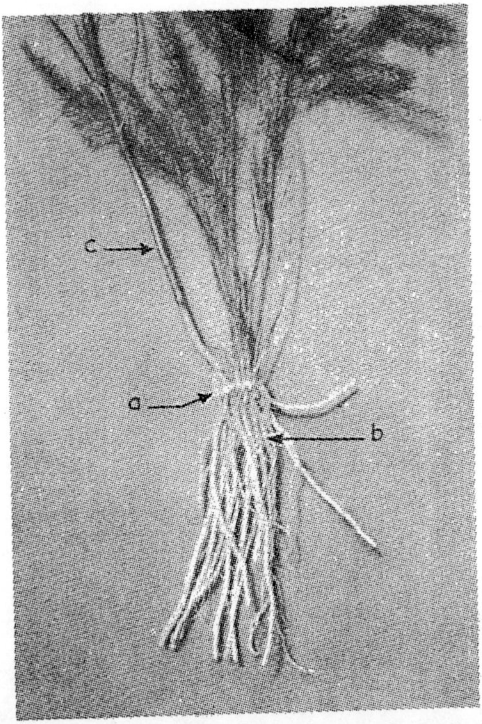

FIG. 62.—Asparagus plant, showing rhizome (*a*), fleshy roots (*b*), and aerial stems (*c*).

troublesome pest and is difficult to eradicate because the rhizomes carry it through unfavorable seasons.

Asparagus is commonly considered as propagated from rhizomes. The underground stems of this plant grow very slowly, only 1 to 2 inches per year. From these rhizomes are produced thick fleshy roots that contain reserve food materials, and they in turn give rise to fibrous branch roots that function as absorptive organs. In propagation of the plant, seed are sown to produce crowns. These consist of a central portion of rhizome and a mass of fleshy roots. When transplanted into their permanent location, during the winter or early spring following the first season of growth, they are referred to as 1-*year crowns*. The rhizome proper is the only portion of the crown that is directly concerned in reproduction of new shoots and fleshy roots.

In German iris (*Iris germanica*) the growth of the rhizome is rather vigorous and the plants spread rapidly. The species of

FIG. 63.—Canna plant, showing rhizome and aerial shoots.

Pogoniris also produce rhizomes that may be divided immediately after flowering for propagation. The canna is another common garden plant that is propagated in similar fashion. Among the fruit plants the banana is one of the most important to be propagated from rhizomes. The roots are dug and cut into "bits" or pieces that weigh from 2 to 4 pounds and contain one or more well-developed buds. This method is used almost entirely in setting new plantings, although "sword suckers" are frequently used for replants.

Although there are still large numbers of the ordinary "flag" iris grown in this country, because of the ease with which they are naturalized, there is considerable interest in the more delicate Palestine iris and its hybrids. Chief among these are the two groups of *Regelia* and *Oncocyclus*, which include some of the most beautiful and interesting types of the entire genus. Although the two species are natives of Palestine, the hybrids have been produced near Haarlem, in the Netherlands. These forms are best propagated by the underground parts, which are neither

Fig. 64.—Bulbs borne terminally on stalk of "tree onion."

rhizomes nor corms but are referred to as *pseudocorms*. The small pseudocorms are united by pieces of rootstock. There is a large, well-developed central bud on each to produce the next season's growth. On the upper side are several smaller buds that give rise to new corms the next season. The main growing point is at the surface of the ground, while the smaller buds are more deeply covered.

Planting of these corms has been done at Chico, Calif., in September in beds 3 feet apart, either in a 1-foot matted row or in

2 rows 6 inches apart. The corms are dug at the end of the second year with a short spade and are stored in a cool, dry place. The better varieties have increased about 3½ times (350 per cent) during this 2-year period.

The production of seed in these hybrid forms has been found extremely variable, both in California and in Virginia. Good moisture conditions are considered essential in production of seed, and it is also thought that careful fertilization will give increased yields. The seed are very slow and uncertain to germinate, and a large number of albinos are produced. For all of these reasons the culture of these forms from seed is very unsatisfactory in this country.

2. *Pip.*—A pip, as produced by Lily of the Valley, is an upright part produced upon a horizontal rootstock. Its origin is thus identical with that of other rhizomes, but it is frequently considered separately.

Tubers.—There are two general classes of tubers; stem tubers and root tubers. The former are characterized by the presence of scale leaves and the absence of a root cap, while the latter have a root cap and no scale leaves. The internal structure of the two is also quite distinct.

1. *Stem Tuber.*—A stem tuber, as typified by the Irish potato, consists of a shortened, thickened, underground stem with scale-like leaves or leaf scars subtending the eyes. The tubers are borne on underground lateral stems or "stolons," which arise at the nodes of the underground portion of the stem. The outer covering of the potato is periderm, in which lenticels replace the stomata. Beneath the surface are found first a layer of cortex, next the vascular cylinder, and in the central portion of the tuber a considerable area of pith. The parenchyma cells of the pith contain some starch but not so much in proportion as the cortex.

Propagation of tuberous plants of this type is accomplished by the formation of new shoots from the modified axillary buds. In most cases, the tuber is cut up into pieces, each of which contains one or more buds and a sufficient amount of stored food for the early growth of the young plant. In the potato and many other tubers the apical buds are dominant, but in some forms the buds near the base give rise to the first shoots.

2. *Root Tuber.*—Fleshy roots produced by plants such as sweet potato, *Dahlia*, and the orchids resemble stem tubers but possess

a rootcap not found in the other group. They do not have the scale leaves and buds found on stem tubers. The old tubers give rise to shoots that in turn produce new tubers from adventitious roots near the base of the shoot, and also from branch roots. Beet and carrot produce a tuberous main root that gives rise to a new top in the second growing season.

In the sweet potato, roots arise on stems from opposite sides of each node. These roots are all fibrous when first developed, but many of them become considerably thickened during the growing season, to form the fleshy, edible portions. The prostrate stems, or vines, will also form roots where they are in contact with moist soil. In commercial practice, the stems or "slips" are produced in hotbeds. They are then removed from the old root and used as cuttings for field planting.

Special Treatments.—Bulbs of various kinds and other subterranean structures are given artificial treatment for two distinct purposes: to destroy insect and disease pests and to break the rest period.

1. *Treatments for Pests.*—The treatment of *Narcissus* bulbs for control of the *eelworm* consists of soaking the bulbs in water at 75°F. for a period of 2 hours and then of holding the bulbs in hot water until all have reached a temperature of 110 to 111°F. They are then held at 110°F. for 4 hours. A pint of formalin may be added for each 35 gallons of water to protect against diseases. Bulbs that have received the hot-water treatment as outlined have made more satisfactory growth than untreated bulbs, but this has been considered a result of their entire freedom from insect pests and not of any physiological change in the bulb.

Mealy bugs are found occasionally on *Narcissus* bulbs. The best treatment for these insects is to dip the bulbs in a solution of soap and tobacco, made with 1 pound of yellow laundry soap, 5 fluid ounces of 40 per cent nicotine sulphate, and 20 gallons of water. Treatment should last only 1½ minutes, and the bulbs should be drained and dried immediately. In severe infestations a second treatment at the end of 10 days is recommended. A similar solution is used to control *aphids* on lily bulbs.

Fumigation is used to destroy larvae of the *bulb fly*. The most satisfactory fumigants are carbon disulphide and paradichlorobenzene. The vacuum method is recommended in connection with the use of carbon disulphide. The cleaned bulbs are placed in a fumigation tank and a 27-inch mercurial vacuum is produced. The dosage of 2½ to 3 pounds to 100 cubic feet of space is then introduced and normal atmospheric pressure is restored. The period of exposure is from 1¾ to 2 hours. The optimum temperature is 70 to 80°F. Paradichlorobenzene, 4 ounces per cubic foot, is placed in the bottom of a fumigation box and covered with burlap. The bulbs are placed on the trays in

FIG. 65.—Equipment used in sterilizing bulbs. (*Courtesy, Bureau of Plant Industry, U.S.D.A.*)

layers 4 inches deep, and the container is closed. Exposure for 120 hours is recommended.

2. *Rest Period.*—The rest period in bulbs is the interval between the time of digging and the time the bulb will resume growth. It may be shortened by conditions of storage. *Narcissus* bulbs are kept in common storage until 4 weeks before planting and then at 50°F. for this last period. Bulbs for early forcing are secured by digging 3 weeks or more in advance of the usual stage and also by growing the stock in a very early region in the season preceding the forcing of the bulbs. The last two

treatments do not shorten the rest period but merely advance its initiation.

Buds of a recently harvested Irish potato tuber will not ordinarily grow for some time after the potato is planted, even under conditions favorable for growth. This time after harvesting during which these buds will not sprout is the rest period. Some writers call it the *dormant period.* Fully matured potatoes have a shorter rest period than those harvested prematurely. Storage at 82 to 86°F. or treatment with certain chemicals, such as ethylene, hastens the ending of the rest period. Table stock potatoes should remain dormant, and hence conditions that make the rest period shorter are avoided in storing them.

Questions

1. What subterranean structures are considered in the classification with bulbs?

2. What are distinguishing features of true bulbs?

3. In what ways does a true bulb resemble a dormant bud in structure?

4. How are new bulbs produced by old bulbs?

5. What is the cycle of development of a narcissus bulb from the time it originates until it produces daughter bulbs?

6. How does the tulip differ from narcissus in its habits of growth?

7. What is a solid bulb? Give examples.

8. How are new corms borne by the old corm?

9. What are the distinguishing features of scaly bulbs? Give examples.

10. Name the ways by which the lily may be reproduced.

11. What are the types of rootstocks?

12. What features suggest that a rhizome is an underground stem? Name some plants that produce rhizomes.

13. What are the two kinds of tubers?

14. How do the sweet potato and the Irish potato differ in structure?

15. What treatments are used for bulbs to destroy insects?

16. What treatments are used to break the rest period of narcissus? Of Irish potato tubers?

Suggested References

1. Brown, T. A.: Flowering Bulb Culture in Florida, *Fla. Agr. Exp. Sta. Bull.* 48, 1928.
2. Griffiths, David: Commercial Dutch Bulb Culture in the United States, *U. S. Dept. Agr., Bull.* 797, 1919.
3. ———: The Production of Narcissus Bulbs, *U. S. Dept. Agr., Bull.* 1270, 1924.
4. ———: The Madonna Lily. *U. S. Dept. Agr., Bull.* 1331, 1925.
5. ——— and E. O. Orpet: The Production of Hybrid Palestine Iris Corms, *U. S. Dept. Agr., Tech. Bull.* 11, 1927.

6. ——— and ———: A Score of Easily-propagated Lilies, *U. S. Dept. Agr., Circ.* 23, 1928.
7. ——— and ———: The Production of Lily Bulbs, *U. S. Dept. Agr., Circ.* 102, 1930.
8. ——— and ———: The Production of Hyacinth Bulbs, *U. S. Dept. Agr., Circ.* 112, 1930.
9. ——— and ———: Daffodils, *U. S. Dept. Agr., Circ.* 122, 1930.
10. ——— and ———: Tulips, *U. S. Dept. Agr., Circ.* 372, 1936.
11. SHIPPY, WILLIAM B.: Factors Affecting Easter Lily Flower Production in Florida, *Fla. Agr. Exp. Sta. Bull.* 312, 1937.

CHAPTER VIII

LAYERAGE

Stems that form roots while still attached to the parent plant are called *layers*, and the practice based on this phenomenon is known as *layerage*. In some plants artificial methods must be employed, while in others root formation occurs naturally. The rooting medium is usually soil, although other materials are used.

Uses.—Layerage is a rather certain method of inducing rooting. Some plants that cannot be started satisfactorily from cuttings can be grown with relative ease from layers. This is true of apples, plums, cherries, and Muscadine grapes. A cutting, having been severed from the plant on which it grew, often does not remain alive until roots are formed. A layer, on the contrary, is supported by the parent plant indefinitely and, in the meantime, it is likely to develop roots.

Many plants produce natural layers freely and thus provide a ready source of new plants. This is true of the raspberry and strawberry and certain forms of the blackberry and dewberry. In these plants the layers are produced by either runners or upright canes that, by arching, come in contact with the ground and develop roots. Other plants produce natural layers from the crown of the plant. The quince and chrysanthemum illustrate this behavior.

On a small scale, layerage may be used to good advantage, for the reason that the layers do not require the close attention as to watering, humidity, and temperature that cuttings require. Roses are frequently grown from layers for this reason.

Objections to layerage are that it is a slow and cumbersome method of propagation; that it may interfere with cultivation; and that parent plants produce a limited number of new plants, so that a great number of stock plants must be provided as a source of layers. Despite these disadvantages layerage is used generally in the propagation of varieties of some plants.

Simple Layers.—Branches that have formed roots at one point only are called *simple* layers. Such layers are made by bending the branches to the ground and covering the portion just below the tip with 3 to 6 inches of soil. This practice is usually carried on in early spring, before growth has started. The tip of the shoot is left exposed, to form leaves and carry on the normal life processes of the plant. The portion of the stem to be covered may be notched or cut in a slanting manner to induce rooting. A simple *tip* layer is one that forms roots very near the tip. The season of the year for making layers varies with the species. With many species the best results are obtained if layers are made in late winter or early spring; with others, late summer and fall

FIG. 66.—Natural tip layer of blackberry, showing growth of shoots at three nodes at *b*; new plants arising as suckers at *a*.

seem to be the best season. The length of time during which layers are allowed to develop before they are severed from the parent plant likewise varies with different species. Many will make sufficient root and top growth during one season to permit them to be transplanted successfully to a new location; others require two seasons to develop a strong root system. The Young berry, blackberry, rose, rhododendron, bush honeysuckle, and raspberry are examples of plants that may be readily propagated by simple layers.

Compound Layers.—Long shoots that are alternately covered and exposed over their entire length are known as *compound* layers. They normally form roots at each point where they are covered and develop new shoots from the portions that are not covered. When they have grown one season or more, the several layers are severed so as to provide a root system on the proximal portion of each layer and a top on the distal portion. The time of the year for making and harvesting compound layers is influenced by several factors. Normally they are made in late winter and early spring. The rooted layers may be harvested later in the same season; or they may be allowed to grow one or two full seasons in order to develop a strong root system. The long

FIG. 67.—Blackberry shoots which have developed from a continuous layer.

canes of the Young berry provide an excellent opportunity for this practice. Compound layerage is adapted to the propagation of the Muscadine grape. The natural production of "rosettes" and roots by the strawberry plant at each second node of the runners illustrates compound layerage.

Trench or Continuous Layerage.—A method known as *trench* or *continuous* layerage consists in covering a branch for its entire length, thus securing plants from all the buds. This method is especially adapted to the propagation of plum and cherry stocks for use in research investigations. One-year-old plants are set in rows 4 feet apart with the individual plants 3 feet apart in the row. The main stem of each plant is set at an angle of 30 to 40 degrees from the horizontal. The trees are then allowed to grow for one season with ordinary cultivation and care. During the next winter a furrow is opened down the row, and the plants, after being headed back, are laid down this furrow and pegged in place, about 2 inches below the level of the soil. Just before growth starts, the pegged plants are covered with a layer of 1 inch of fine soil. The new shoots push upward through this layer and are etiolated, or white, at the base. As the shoots elongate, more soil is added around them until they are covered to a depth of 5 to 6 inches. If the plants are covered too deeply, the buds will not grow; on the other hand, if the covering is not deep enough, the shoots will not be etiolated.

In the case of layered apples it has been found that etiolation is not essential. The young shoots from the pegged plants are allowed to grow to a height of 5 to 6 inches, and the mounding is then done in one operation. The same procedure as given for plum and cherry stocks is frequently used, however, to encourage early rooting.

In the following winter the rooted plants are severed from the parent stem, which is uncovered and left exposed until new shoot growth begins. The process may be repeated in this manner for several successive years. Some of the more vigorous new shoots fail to root each year, and these may be laid down to supplement the original plant.

Mound or Stool Layerage.—This method is especially satisfactory for the rooting of apple and quince stocks and is used in preference to trench layerage when possible, as it involves less trouble and expense. A stock bed is established by setting young plants 2 feet apart in rows 3½ feet apart. The plants are headed back before growth starts and are allowed to grow for one season. The following winter the plants are cut back within 2 inches of the ground level, with the result that many new shoots arise from the base during the following season.

In the case of apples, which root freely from these new shoots, the stools are allowed to remain uncovered during the early part of the growing season. The greatest number of shoots is produced in this way; after they are formed and have reached the height of 8 inches they are mounded with about 5 to 6 inches of

FIG. 68.—Lowering mound from mother plantation of clonal apples so as to expose rooted shoots for removal. (*Courtesy, H. B. Tukey, N.Y.* (*Geneva*) *Agr. Exp. Sta.*)

FIG. 69.—Removing rooted shoots from layered apple plant. Mother plant may be seen exposed in lower right-hand corner, with rooted shoots arising from it. (*Courtesy, H. B. Tukey, N.Y.* (*Geneva*) *Agr. Exp. Sta.*)

soil. Mounding should be done with moist soil, which should be placed from the center outward, in order to bend the shoots out and give them better spacing. This spacing seems to give a better rooting, especially with vigorous shoots. The tips of the young shoots should never be covered, since this is likely to cause decay.

FIG. 70.—Six rooted apple shoots removed from a layered plant. (*Courtesy, H. B. Tukey, N.Y. (Geneva) Agr. Exp. Sta.*)

When plums are being grown, the procedure is modified and the plants mounded before the new shoots appear. This practice results in the formation of fewer new shoots than the other method, but the shoots that are produced are etiolated and form roots better than those that are produced before mounding. This applies not only to shoots from stools and layers but also to stems used for cuttings, from which better rooting is obtained when their bases have been etiolated during growth.

After the plants have been mounded by either method, they are allowed to grow during the rest of the season, and roots will be formed along the covered portions of the stems on the new shoots. In early winter the rooted shoots are removed and planted in the nursery row. Apples and pears are usually set 1 foot apart in rows 3 feet apart, while plum and cherry stocks are

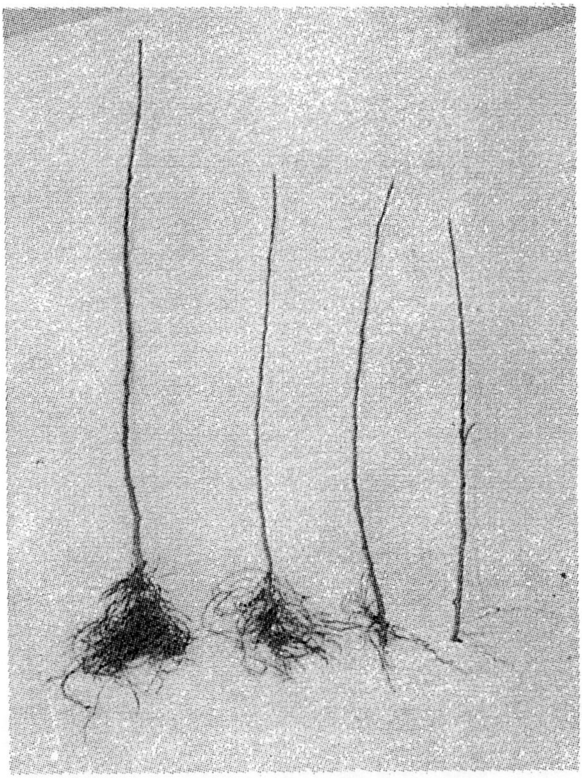

FIG. 71.—Rooted shoots of layered apple, separated and ready to be used as lining-out stock. (*Courtesy, H. B. Tukey, N.Y. (Geneva) Agr. Exp. Sta.*)

planted 1 foot apart in rows 4 feet apart. These plants are set at a depth of about 6 inches. They will be ready to bud in July of the following year, or they may be grafted at the end of one season in the nursery.

The chrysanthemum forms natural mound layers from the overwintering crown at the beginning of each new growing season. These develop into new plants when they are detached and planted out separately. Quince and Japanese flowering quince have habits of growth that permit them to be propagated from natural layers from the crown of the plant.

Air Layerage.—A method used to root branches of stiff-growing plants that do not sprout or sucker readily is known as *air* layerage. Chinese layerage and pot layerage are other names for the same method. The principle involved is to injure the stem by notching or girdling and then to enclose this portion of the

FIG. 72.—Chrysanthemum plant, showing formation of new plants as natural layers from the crown of an old plant.

stem in a suitable medium for rooting. In the greenhouse this may frequently be done by tying a bunch of sphagnum moss around the stem at the desired place and keeping this moss damp by sprinkling when the plants are watered. This method is used on *Ficus elastica*, the ornamental rubber plant, and others of similar nature.

There are many types of pots and boxes that are used in this method, the container being modified in some way to allow the branch to pass through it. Clay pots with a notch in one side are used, as are also metal pots or cones that are hinged on one side. With all these containers the branch is notched or girdled and then placed in the box or pot, which is later filled with soil or some other suitable medium. Suitable supports must be provided to hold up these heavy containers until rooting has taken place. The stem is then severed just below the pot, which will continue to support the rooted plant until it is transplanted.

FIG. 73.—Air layerage as used on plants with stiff stems.

In propagation work by workers of the United States Bureau of Plant Industry, use is made of a "marcottage" box for this type of propagation. This box is essentially a paraffin-coated paper, fiber, or metal box, the lid and bottom part of which telescope together. The two ends of the box are notched or slotted so as to permit the stem of the plant to be encircled closely, but without constriction, when the two sections are brought together. The box is filled with moss or other suitable rooting material of the proper moisture content. The loss of moisture is so slight that such a box needs only infrequent attention, at intervals dependent somewhat on atmospheric conditions.

The nurse-grafted Y-cutting method is an ingenious combination of grafting and air layerage described by the same investigators for the propagation of difficult species. The first step in this process consists in making an approach graft between a stock plant, which is in a pot, and one branch of a Y crotch on the parent plant. As soon as the union is accomplished the top of the stock plant is removed just above the union. Then the portion of the stem on the parent plant, several inches below the Y, is girdled and an air layer is made with the use of the marcottage box previously described. The water and nutrients move up from the root of the stock plant, then downward through the Y and back up into the other branch to keep it growing until the layered base produces roots. When this is finally accomplished, there is one grafted plant and one rooted layer from the parent plant.

Layerage as a Preliminary Treatment.—Layerage is sometimes used as a preliminary treatment for the rooting of cuttings. It frequently happens that the layered parts do not form satisfactory root systems in the first season, but the stem pieces with small roots may be separated from the plant and treated as cuttings. Another use of preliminary layerage is for the purpose of etiolating the stem so as to induce rooting by cuttings. Work done by several investigators in England shows that the etiolated portion of the young stem forms a superficial starch sheath from which roots may be produced more readily.

Questions

1. What is layerage?

2. Under what conditions is it used in preference to some other method?

3. What are the objectionable features of layerage?

4. List the different types of layers. Prepare sketches illustrating each type.

5. What is meant by etiolation?

6. What is the value of notching or girdling as a preliminary treatment for layers?

7. Layerage is used as a preliminary treatment for material to be used for what purpose? Explain.

Suggested References

1. GARDNER, F. E.: The Vegetative Propagation of Plants, *Md. Agr. Exp. Sta. Bull.* 335, 1932.
2. HANSEN, C. J., and E. R. EGGERS: Propagation of Fruit Plants, *Calif. Agr. Ext. Serv. Circ.* 96, 1936.
3. KNIGHT, R. C., J. AMOS, R. G. HATTON, and A. W. WITT: The Vegetative Propagation of Fruit-tree Root Stocks, *East Malling Research Sta.* (Kent, Eng.) *Ann. Rept.* (14th and 15th years), II, Supplement, pp. 18–19, 1928.
4. STAHL, J. L.: Propagation of Deciduous Fruits, *Calif. Agr. Exp. Sta. Circ.* 294, 1925.
5. SWINGLE, W. T., T. R. ROBINSON, and EUGENE MAY: The Nurse-Grafted Y-Cutting Method of Plant Propagation, *Jour. Hered.*, **20**: 84, 1929.
6. TALBERT, J. T.: Plant Propagation by Seedage, Cuttage, Layerage, and Separation, *Mo. Agr. Exp. Sta. Circ.* 191, 1936.
7. YERKES, GUY C.: Propagation of Trees and Shrubs, *U. S. Dept. Agr., Farmers' Bull.* 1567, 1929.

CHAPTER IX

CUTTAGE

Cuttage is the process of propagating plants by the use of vegetative parts that, when placed under suitable conditions, will develop into complete plants. It differs from layerage in that the parts used are detached from the parent plant before they have an opportunity to develop roots. With species of plants that strike roots readily, cuttage is a cheap and convenient mode of propagation. It is used extensively in the propagation of ornamental plants, including deciduous types, broad-leaved evergreens, and coniferous forms. Some fruits, such as grapes and figs, have been propagated in this manner since ancient times, and more recently there has been considerable progress in the rooting of other fruit plants. Some varieties, such as the Bruce plum, are grown on their own roots. In the majority of cases, however, the rooting of fruit-tree species is of more importance in the production of uniform stocks for budding or grafting.

CLASSES OF CUTTINGS

Plant parts used in making cuttings fall into four groups: roots, leaves, stems, and modified stems (tubers, rhizomes, and similar structures). Theoretically, all plants that have primary meristems are capable of being propagated by cuttings. All plants cannot profitably be increased by this means, however, and only practical experience has made it possible to distinguish between species that can be propagated from cuttings and those that cannot.

Root Cuttings.—As a rule, plants that naturally produce suckers freely can be propagated easily by root cuttings. Some species of plants that root rarely or not at all from stem cuttings can be reproduced by means of root cuttings. Persimmon, pear, pecan, apple, and plum are of this class. They may be started by root cuttings, but other methods are considered more economical and are in general use. Sweet potato and horse-radish are propagated commercially by root cuttings, and blackberries and raspberries may be propagated successfully by this method. It should be borne in mind, however, that a root cutting will perpetuate the part of the plant from which it was secured. A root cutting taken from below the union of a budded or grafted tree

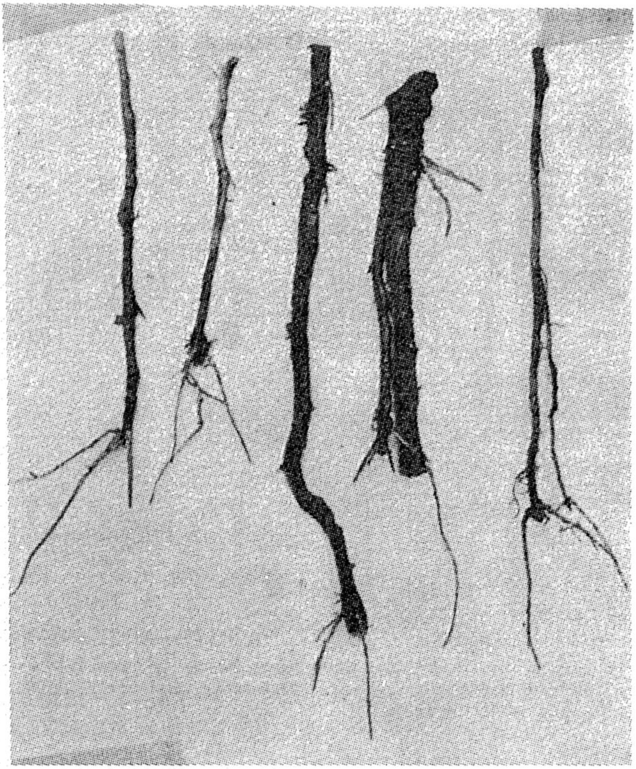

Fig. 74.—New roots formed on root cuttings of pecan.

reproduces the seedling stock of unknown bearing quality rather than the standard top.

The technique of making root cuttings varies widely with different species. They are customarily made from roots that are not smaller than ¼ inch in diameter, which are cut in lengths of 2 to 6 inches. They may be made early in winter, stored in sand, and allowed to callus. They are then planted out in the open the following spring. By another practice the cuttings are started in early winter in greenhouses or hotbeds and transplanted to the open after they have made top growth and formed new roots; such plants are usually large enough to be transplanted by spring. Root cuttings are also planted directly in the field in the spring without preliminary treatment. They may be planted in either a horizontal or vertical position; if planted vertically, the end that was nearest the crown of the parent plant should be uppermost. New shoots develop from root

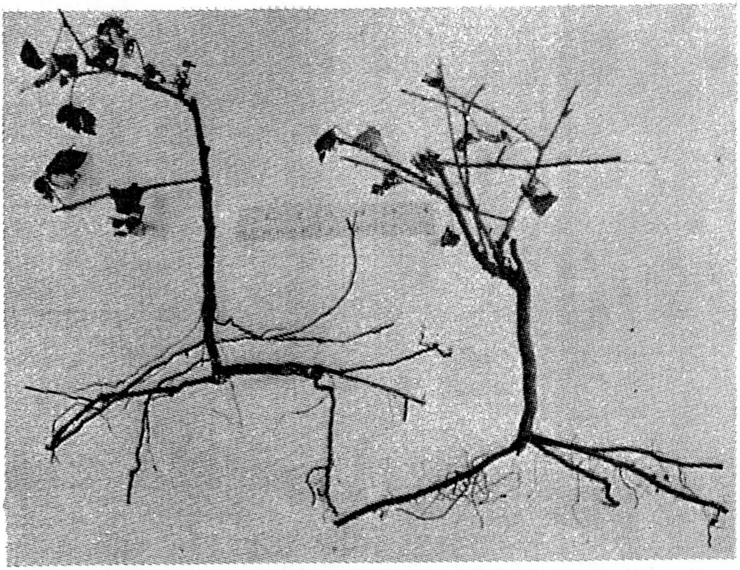

FIG. 75.—Blackberry plants that have grown one season from root cuttings; one on left is from a root cutting planted horizontally and one on right is from root cutting planted vertically.

cuttings from adventitious buds, and new branch roots form either from the old root part used as a cutting or from the base of new shoots that develop from below ground.

Leaf Cuttings.—Many tropical plants with leaves having fleshy, radial veins are propagated by leaf cuttings, which are cut with a vein in the center and a portion of the petiole or leafstalk at the base. The roots and the shoot both develop at the base of the cutting, and the original leaf seldom if ever becomes a part of the new plant. Another practice is to cut the leaf transversely across each vein, lay it flat on sand, and cover each cut.

It is best to use rather mature leaves in either case and to give them bottom heat.

This method of propagation is used to a considerable extent with various greenhouse plants used for ornamentals. *Peperomia* will root from a cutting of the entire leaf and half of the petiole, the latter being inserted in the sand. Snake plant, which has a long leaf, will root from each piece if the leaf is cut transversely. In *Chrysanthemum* it is best to use leaf cuttings with a portion of woody stem at the base of each. The leaf of the rubber plant (*Ficus elastica*) will root but does not produce a plant unless a portion of the stem is attached. *Gloxinia* and *Bryophyllum* will root from leaf cuttings, and leaves of the lemon (*Citrus limonia*) and orange (*C. sinensis*) may be rooted in the same manner.

Stem Cuttings.—These are made from herbaceous plants, such as those frequently grown in greenhouses, and from woody plants, which are usually grown in the open. Cuttings of woody plants may be classed as semihardwood, or softwood, and hardwood, depending upon the stage of growth of the wood used.

1. *Herbaceous Cuttings.*—These are made mostly of greenhouse plants that are herbaceous in type. Cuttings of such material are usually soft, tender, and succulent; they require special attention with regard to temperature and moisture to prevent wilting. Under favorable conditions they root satisfactorily in a relatively short time. Examples of plants that may be propagated by herbaceous cuttings are *Geranium, Coleus, Petunia, Alternanthera,* Wandering Jew, *Chrysanthemum,* tomato, and sweet potato.

2. *Semihardwood Cuttings.*—Stem cuttings of trees and shrubs that are made from current-season shoots are known as *semihardwood,* or *softwood,* cuttings. In practice they are made 3 or 6 inches long. Cuttings that are made so as to include terminals of growing shoots are generally preferred, though those made from parts below the terminal are satisfactory. Shoots that snap clean when broken are considered to be in ideal condition for use as semihardwood cuttings. The leaves are removed from the basal portion, but those near the terminal are left.

Semihardwood cuttings are succulent and tender; for this reason it is important that they be handled so as to prevent wilting after they are cut and before they are planted. The presence of leaves causes a high rate of transpiration, which makes this difficult. Best results may be secured by cutting them during a cool part of the day, preferably in the early morning while the material is turgid. They should then be wrapped in moist cloth or moss until planted. Such cuttings are usually started in specially prepared beds in a greenhouse, hotbed, or cold frame. The solar propagating frame and the box frame have been used successfully for the cuttings of some plants; others, such as rose and lantana are sometimes started outdoors.

In addition to cool temperature, shade, and a high humidity, which are essential factors for good results with semihardwood cuttings, bottom heat may also be supplied in order to provide more desirable conditions for rooting. Manure is frequently used for this purpose, or the beds may be heated with flues, hot water, or electric heating elements. Shade may be provided by stretching domestic cloth at a height of 3 or 4 feet above the bed, or the glass of the greenhouse may be sprayed with lime whitewash to provide the same effect. On a small scale, cuttings may be planted in shallow boxes or flats placed in a shaded location. The cuttings and adjacent areas are sprayed with water several times a day to keep the cuttings from wilting.

3. *Hardwood Cuttings.*—These are made from a wide variety of plants, including deciduous types, conifers, and broad-leaved evergreens.

a. Cuttings of *deciduous* plants are taken during the dormant season. Those of some plants are taken in the fall, packed in moist insulating material, and stored at a temperature of 40°F. or less. These cuttings are usually placed in the bed about midwinter. While in storage they may have formed callus at each end; this, however, is not essential to rooting. Instead of the procedure just outlined, cuttings of some deciduous plants are taken and planted in late winter, shortly before the resumption of growth.

Deciduous cuttings may be made from 4 to 12 inches long, depending on the kind of plant. Usually one-year-old wood is used, but the older wood also may be rooted. It is a customary practice to make the top cut slightly above a node, and the lower cut slightly below a node. Various kinds of cuttings show different responses with regard to the point of origin of roots;

but the denser tissue in the vicinity of the node is thought to be of value in preventing drying out or decay of the wood. Deciduous hardwood cuttings are not highly perishable, but they should be protected at all times to prevent them from becoming dry.

Many species of plants may be propagated by hardwood cuttings set directly in the nursery row. Grape, fig, and rose

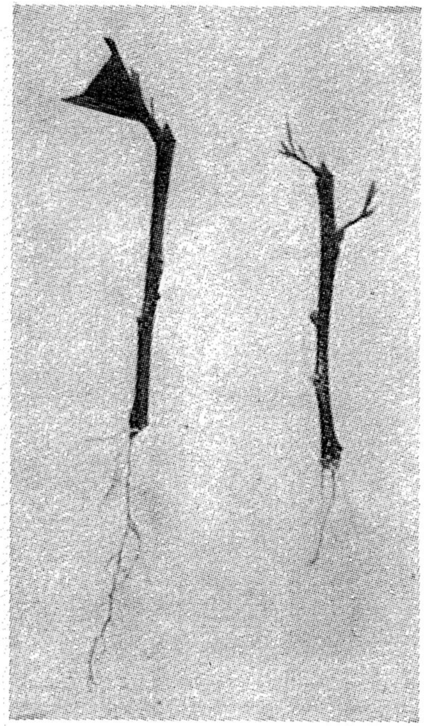

FIG. 76.—Cuttings of wax-leaf Ligustrum showing influence of old leaf on root growth. The full-length root at the left shows the difference in response.

are commonly propagated in this manner. Rooting is determined partly by the type of soil in which they are planted; sandy loam soil that is well-drained is preferred. In order to insure good aeration, cuttings are frequently planted on high beds. In a heavy clay soil in Oregon, grape cuttings rooted well when set in holes made with an iron bar and filled with sand.

b. Hardwood cuttings include also those made from mature wood of *conifers*. Cuttings of such plants are made 4 to 6 inches long with foliage removed from the lower portion of the stem. As the cuttings form roots, new shoots form also, and this top growth is an indication that the cutting is ready to be moved. The customary procedure is to pot the rooted plants and grow them in the pots for one season before moving them to the field.

FIG. 77.—Cuttings of French crab seedlings. 1 and 2, in sand, unrooted; 3 and 4, in sand and peat, early in season—basal roots; 5 and 6, in sand and peat, later in season—basal and nodal roots.

Some of the arborvitaes root within 2 or 3 months; junipers frequently require 6 months or even longer.

c. Several *broad-leaved evergreen* plants are grown from hardwood cuttings. The cuttings of certain citrus species, for example, are made 4 to 7 inches long with five to six nodes from mature terminal growth. Leaves are removed from the lower part of the stem, but two or more are left at the top. As with other types of cuttings, it is important that cutting material be obtained from healthy, vigorous-growing trees. Oranges, grapefruit, lemon, American holly, yaupon, and several species of *Ligustrum* are examples of broad-leaved evergreens that may be propagated by hardwood cuttings.

Origin of Roots in Hardwood Cuttings.—It has been a rather common idea that the stem cuttings of different species have a definite manner of rooting, some of them at the base only, others at nodes along the stem, and still others at internodes. It is known that roots which develop on hardwood cuttings originate in two different ways. A recognition of the difference between these two types of root formation may be of value in explaining the variable rooting responses of some plants.

FIG. 78.—One-year stem of Chenango apple with roots at basal end.

1. *Adventitious roots* develop usually in the cambium but in some cases from the callus. They usually form just prior to their emergence. This type of root formation is probably of most general occurrence because these roots may be formed at almost any time of the year. They frequently develop in the immediate vicinity of the basal cut, and because of this they are sometimes known as *wound roots*. Adventitious roots, however, may occur at other points on the portion of the stem of the cutting that is in contact with the rooting media. They may develop at either the node or internode.

2. *Preformed root initials*, known as "root primordia," occur in the stems of certain plants. These appear as small groups of cells or very small roots, clearly differentiated in the cambium region, and are symmetrical arrangements of meristematic tissue. They develop until the tips reach the outer surface of the bark and then cease growth and become dormant. When hardwood cuttings are made from stems containing such primordia, roots form freely from the primordia. In one variety of apple and in the willow,

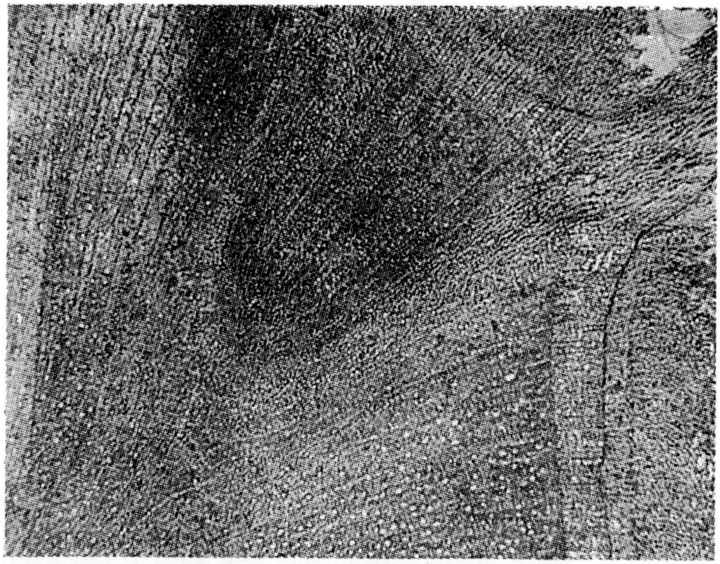

Fig. 79.—Section through base of cutting showing root origin in wood. (See Fig. 78.)

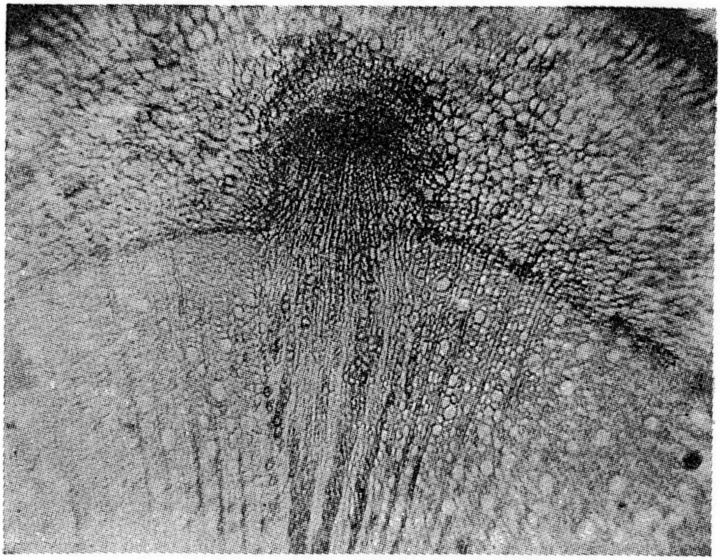

Fig. 80.—Root primordia in quince stem.

preformed roots seem to be formed only toward the end of the dormant season. Cuttings of these plants taken early in the dormant period form roots only at the base; those taken in December or later form roots at the nodes along the stem as well as at the base.

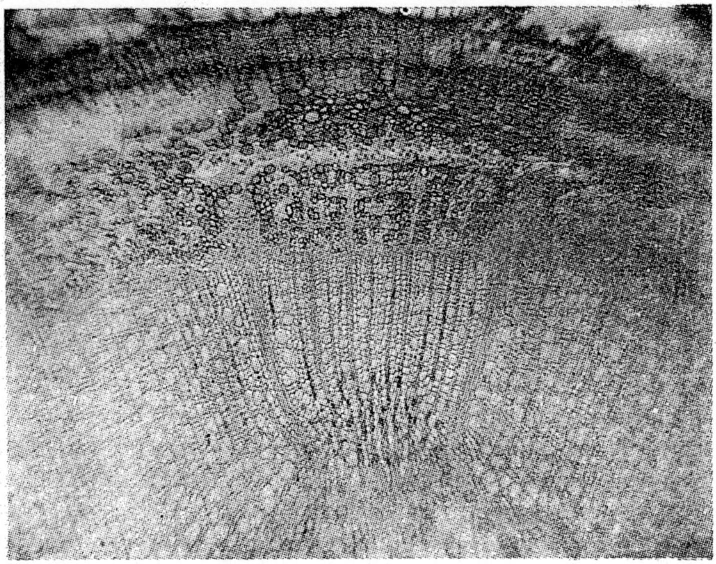

FIG. 81.—Root primordia in willow stem.

Cuttings of Modified Stems.—Material of this type is usually handled as bulbs and bulb-like structures, and it has accordingly been discussed in connection with bulbs.

FACTORS INFLUENCING ROOT FORMATION IN CUTTINGS

EXTERNAL OR ENVIRONMENTAL FACTORS

External or environmental factors represent the treatments that are applied just before the cuttings are set in the bed, or the conditions to which the cuttings are subjected in the bed.

Media.—Several different materials are used for the rooting of cuttings.

1. *Sand.*—Clean, sharp sand is used more nearly universally than any other material. It should be sufficiently fine in structure to be retentive of moisture but at the same time must provide adequate drainage. Ordinary building sand will frequently be

found satisfactory for this type of work. If the sand is of varying physical composition, containing lumps of clay, gravel, or organic matter, it should be screened before it is used. Decaying organic matter in the sand is very objectionable. Such material promotes the development of fungi and bacteria, which in turn may cause the cuttings to die before root formation can take place.

Sand used in the cutting bed should be changed regularly after the plants are removed or should be sterilized effectively before being used again. This is especially important because in the transplanting of rooted cuttings a considerable portion of the root system may be broken and left in the soil. Dead cuttings are also frequently left in the bed too long, so that ready sources of infection are provided for the next lot of cuttings.

2. *Acid Peat.*—In comparatively recent years, acid peat has been used successfully as a rooting medium for cuttings. Peat is composed largely of partly decomposed organic material. It is normally brown in color, light and granular in texture, and acid in its reaction. It has a high water-holding capacity; saturated peat contains over three times its weight of water but is well aerated even when saturated. Aeration and retention of moisture are two important requirements of a rooting medium. The acid reaction of the peat is considered to be beneficial or even necessary for some cuttings that root more satisfactorily in it than in sand; it serves also to prevent bacterial decomposition. This seems to be true for arborvitaes, junipers, holly, and Florida dogwood.

3. *Sand and Peat Mixture.*—Cuttings of plants that root poorly when placed in sand often root satisfactorily in a mixture of equal volumes of sand and peat. The superiority of this medium over sand for certain species is probably due to improved aeration and increased water-holding capacity. The mixture influences not only the rooting but the position of roots that form, tending to cause them to be massed at the base of the cutting in some species.

4. *Other Media.*—Various other materials, such as coconut fiber, rice ashes, and sphagnum moss, have been used successfully in cutting beds. Alone or in mixture with sand they provide media that are well-aerated, well-drained, and yet retentive of moisture.

Temperature.—Control of temperature is a very important factor in the rooting of cuttings that are somewhat difficult to start. In the case of hardwood cuttings, placed in the bed in winter or early spring, the primary consideration is to induce root activity before shoot growth occurs. For this reason, it is desirable to provide bottom heat, so that the bed itself is 5 to 10°F. warmer than the surrounding air. This is accomplished by placing heating pipes below the surface of the rooting medium in the bed, or by placing them entirely underneath the raised bed, and boxing up the space on the sides. Root formation may occur over a wide range of temperature, but a soil temperature of 65 to 70°F. gives satisfactory results with many plants. In recent years excellent results have been obtained in the heating of hotbeds and greenhouse benches by electric heating elements, which give a very uniform temperature.

In certain cases, cuttings of coniferous evergreens are rooted in the summer in a closed greenhouse. Even where the house is whitewashed or shaded, the temperature rises to a very high degree. Careful attention to watering is necessary in such cases to prevent the material from burning.

Humidity.—A high degree of humidity should be maintained in the cutting bed in order to prevent drying and death of the cuttings before they have opportunity to root. This is especially important for herbaceous, softwood, and evergreen cuttings. Frequent sprinkling of walls, walks, and beds in the greenhouse is necessary in order to keep the cuttings from drying and wilting. In hotbeds, cold frames, or other propagating structures, glass sash or other types of covers are used to prevent water loss. Under such conditions of high humidity, diseases of all kinds find a favorable condition for rapid spread. Careful sanitation of the cutting bed is one of the essential considerations to keep down such diseases.

Chemical Treatments.—Many kinds of chemicals have been used to induce root formation in species difficult to propagate or to increase the number and extent of roots in others that develop slowly. Early work with dilute solutions of potassium permanganate (0.01 per cent to 0.001 per cent) on privet cuttings called attention to this method as a possible aid in stimulating root formation. Dilute solutions of vinegar (0.01 per cent acetic acid) and cane sugar (2 per cent) also encourage rooting of cuttings of some plants. In all these treatments the solutions are placed in beakers in which the cuttings are standing vertically with the basal ends down. The solution should cover the lower third of the cuttings, which are subjected to the treatment for 12 to 24 hours before being placed in the beds. Each of these treatments has given favorable results with some plants, but in some cases the percentage of rooting is not increased above that of the check plots and in other cases rooting is definitely retarded or entirely prevented.

FIG. 82.—Camellia cuttings. Photograph shows characteristic response of those which are not treated in comparison with those which are treated with Hormodin, a preparation of indolebutyric acid. Picture shows the cuttings as they appeared 60 days after they were placed in the rooting medium. (*Courtesy Hitchcock and Zimmerman, Boyce Thompson Institute for Plant Research, Inc.*)

A recent development that has attracted considerable interest is the use of plant hormones and synthetic growth substances to stimulate plant growth and especially root formation in cuttings. The first of these hormones were derived from plant tissues; *auxin a*, *auxin b*, and *heteroauxin* (β-indoleacetic acid) were among the first of these substances to be used. More recently, workers have discovered that several crystalline acids and also several gases produce similar root stimulation. The specific compounds that have been used successfully include the following crystalline acids: indoleacetic, indolebutyric, indoleproprionic, phenylacetic, and napthalenacetic. These materials are dissolved in small amounts of alcohol and diluted with tap water. The basal ends of the cuttings are placed in the solutions to a depth of ½ to ¾ inch. Concentrations used range from 1 to 40 milligrams in 100 cubic centimeters, and the duration of treatment should be inversely proportional to the concentration; it may vary from 6 to 96 hours. Results that have been obtained so far show that several of these chemicals are of value in stimulating root formation in some species. In other trials, it has been

FIG. 83.—Carnation cuttings, photographed three weeks after planting in the cutting bed. Left, planted without treatment; right, dipped in Hormodin powder before being planted. (*Courtesy Hitchcock and Zimmerman, Boyce Thompson Institute for Plant Research, Inc.*)

shown that root development is initiated by gases, such as ethylene, acetylene, propylene, and carbon monoxide.

Although the hormone treatments are promising, they have not been uniformly effective in causing difficult plants to root. Hardwood cuttings of apple, black walnut, chestnut, and many other species have not made a satisfactory response to the customary treatments. Recent investigations have shown that shipmast locust and pecan will form roots, following treatment with solutions of 80 or 160 milligrams of indolebutyric acid per liter, instead of the more common treatment of 20 milligrams per liter. Injury to the basal ends of cuttings, which was observed in fresh material, was overcome by causing the cuttings to callus before treatment with the chemical.

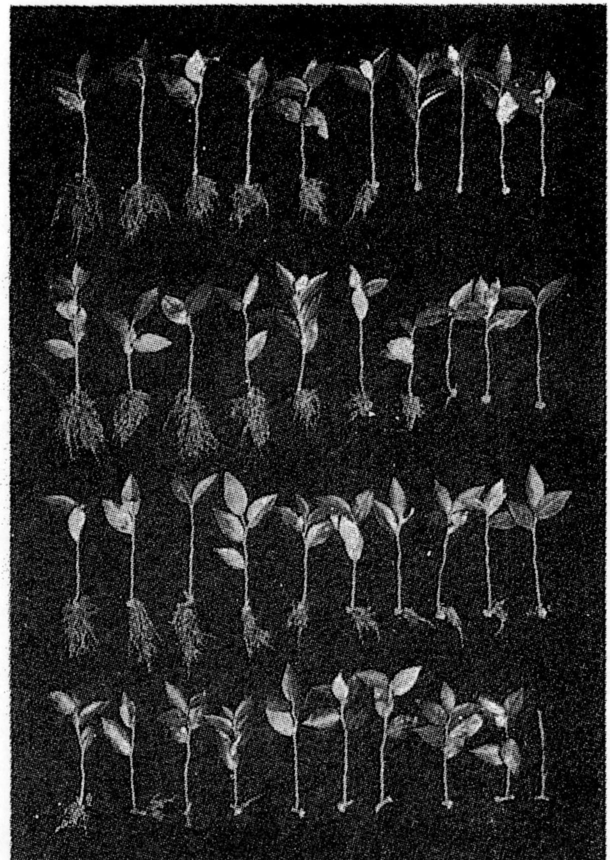

FIG. 84.—The difference in rooting of highbush blueberry (*Vaccinium corymbosum*) from four types of softwood cuttings. Top, cut made above base; second row, cut made at base; third row, heel cuttings; bottom row, mallet cuttings. (*Courtesy Hitchcock and Zimmerman, Boyce Thompson Institute for Plant Research, Inc.*)

Mechanical Treatments.—Mechanical treatments of various kinds have been used to stimulate root formation. Some of these treatments are used on the plant before the cuttings are made, with the result that internal or structural changes are induced. For this reason they are included under the internal factors.

1. *Retention of Leaves.*—The presence of leaves has a marked effect on rooting of evergreen and also of softwood cuttings.

FIG. 85.—The difference in rooting of a species of plum (*Prunus tomentosa*) from four types of softwood cuttings. Top row, cut made above base; second row, cut made at base; third row, heel cuttings; bottom row, mallet cuttings. (*Courtesy Hitchcock and Zimmerman, Boyce Thompson Institute for Plant Research, Inc.*)

In the case of citrus cuttings the removal of the terminal half of each leaf retards root formation and reduces the total amount of roots produced. Cuttings of American holly and similar species are always made with the upper leaves intact. Most investigators and commercial propagators advise special care of softwood cuttings to maintain their leaves in growing condi-

Fig. 86.—The difference in rooting of American Pillar rose from three different types of cuttings. Top row, cut made above base of shoot; middle row, cut made so as to include a bud of last season's wood; bottom row, mallet cuttings. (*Courtesy Hitchcock and Zimmerman, Boyce Thompson Institute for Plant Research, Inc.*)

tion. Such cuttings are said to root sooner and in greater percentage when the leaves are retained.

2. *Position of the Basal Cut.*—It is frequently stated that the basal cut should be made through or just below a node. Many

species do form roots more readily, and some of them exclusively, at the nodes. In such plants it would be advisable to pay more attention to the basal cut. Many other species, however, root

FIG. 87.—The difference in rooting of Dorothy Perkins rose from three different types of cuttings. Top row, cut made above base of shoot; middle row, cut made so as to include a bud of last season's wood; bottom row, mallet cuttings. (*Courtesy Hitchcock and Zimmerman, Boyce Thompson Institute for Plant Research, Inc.*)

freely along the internode or at the basal end, whether this happens to be a node or internode.

3. *Slope of the Basal Cut.*—The base of the cutting should be cut perpendicular to the main axis in species that root at the base or are difficult to root. It has been observed that more

roots are formed in this case than in other cases where a slanting cut is made. In species that root freely, little attention is paid to the slope of the cut.

4. *Notching or Splitting the Basal End.*—In some of the older treatises the recommendation is made that the base of the cutting should be split and a small pebble inserted to hold the split end open. The practice is seldom carried to that extreme, though some cuttings respond favorably to notching and splitting.

5. *Type of Wood.*—The type of wood at the base has an important influence on the rooting of cuttings of some plants. Cuttings may be made of current-season growth or green wood, with the basal cut made in one of the following ways: (*a*) cut made slightly above the base of the current season's growth; (*b*) cut made at the base of current season's growth; (*c*) a heel of one-year wood; (*d*) a mallet of one-year wood.

Internal or Structural Factors

Internal or structural factors represent conditions within the cutting, which may influence its response. Such conditions may be affected by treatments to which the cuttings are subjected some time before they are removed from the plant. These treatments differ from external treatments in that they are designed to induce some change in the chemical composition or structure of the material before it is planted in the cutting bed.

Stored Food.—Stored food may have considerable influence on the rooting of cuttings. In California, cuttings of Sultanina grape were sorted into three classes on the basis of their starch content. The freshly cut ends of the cuttings were dipped in a solution of iodine in potassium iodide, and the intensity of the staining in wood outside the medullary rays was used as the standard for selection. Cuttings that showed the deepest stain, presumably indicating the highest starch content, rooted 62 per cent, the intermediate group 35 per cent, and the low-starch group 17 per cent. The difference in vigor of the root growth in the various groups was even more pronounced.

Shoots of stock plants from which cuttings are to be taken are sometimes girdled in order to influence the amount of stored food that the cuttings will contain. The girdle is made at the point on the stem which will be the base of the cutting. The resulting swelling above the girdle is accompanied by an accumulation of stored food at this point. The girdling of the shoots is done during the growing season as soon as length growth ceases, and the material is then removed for cuttings during the following dormant period. The additional amount of reserve food accumulated at the base of the cutting is presumably of value in promoting root formation; but the method would not be practiced except in cases of plants that are difficult to propagate. A further modification is to cover the girdled area, and the stem just above it, with damp moss or with soil in a pot. Root

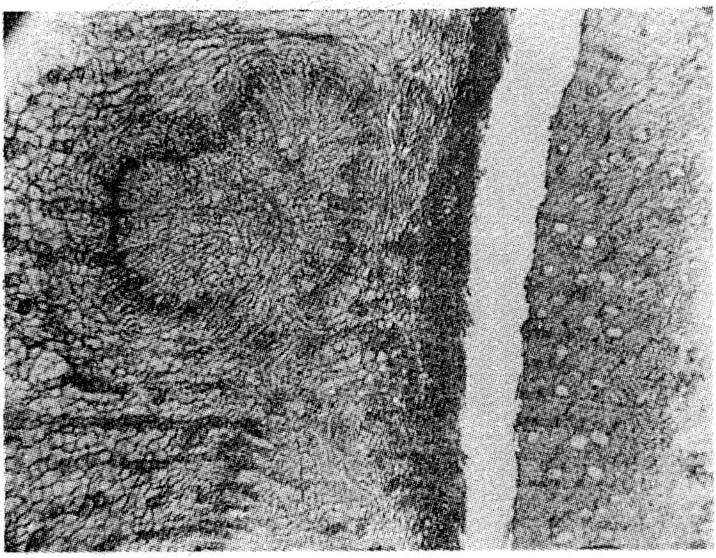

FIG. 88.—Section showing root origin in callus at base of cutting.

formation then occurs before the shoot is removed, the method thus becoming air layerage.

Age of Wood.—The age of the wood is usually not considered, since most cuttings are made of wood one year old, or of current-season shoots. Some plants will root readily from older wood. Recent work with pecan cuttings, as previously described, was done with wood two to four years old. In Italy the olive is propagated by means of "truncheons" or pieces of old wood. The age of the *plant* has been found to be a factor in rooting of some species. In Maryland it was found that cuttings of one-year seedlings of apple, peach, and other fruit trees would form roots; but one-year wood on trees two years old or older would not give comparable results. When the plants were cut back to the crown, the new shoots gave a similar response to that of the one-year seedlings. No anatomical differences to account for this variation in behavior were mentioned.

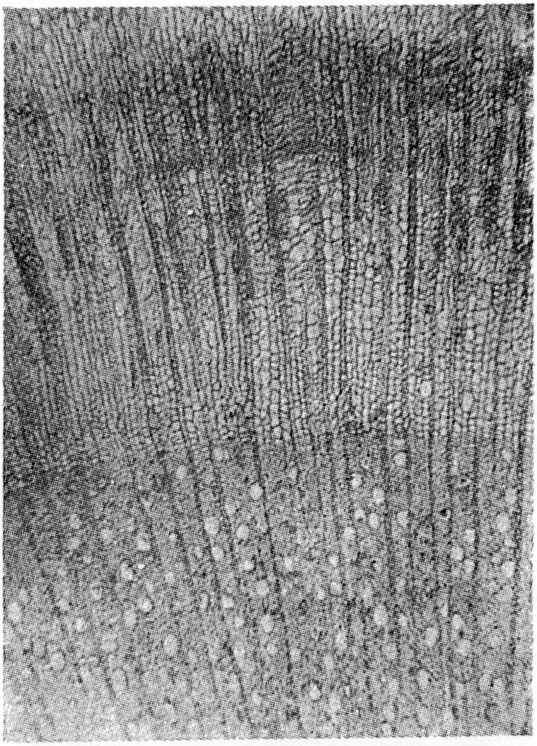

FIG. 89.—Section slightly above that shown in Fig. 88 showing no vascular connection between root and stem of cutting.

Callus Formation.—Callus formation at the basal end of the cutting was at one time considered to be a vital factor in the rooting of hardwood cuttings. More recently it has been accepted that it does not play an important part in root formation, and that in many cases the callus interferes with the emergence of roots. Some types of cuttings that form large masses of callus tissue are removed from the bed and the callus trimmed away in part to facilitate rooting. Some few cases have been observed where roots originated in the callus tissue, but that is uncommon. Callus formation may be of benefit in sealing the end of the cutting and preventing decay. Callused cuttings may also be treated with higher concentrations of chemicals to aid in root formation than those not callused.

FIG. 90.—Callus and root formation on a cutting. (*Courtesy Homer E. Rea, Tex. Agr. Exp. Sta.*)

Etiolation.—Parts of shoots not containing chlorophyll are said to be *etiolated.* The etiolated condition is regarded as being favorable to root formation. Etiolation may be produced by wrapping stems with tape or by covering with soil. The exclusion of light causes chlorophyll to disappear. In some cases shoots are caused to develop in darkness, by mounding with dirt, and chlorophyll never develops. Stems arising from below the ground, as in mound or continuous layerage, are etiolated. The anatomical and other changes produced by this process have been interpreted in several different ways. Some investigators state that an endodermis is formed, as in the root, and that root formation is thus encouraged. Others have found that treatment producing etiolation stimulated root formation; but no modifications in structure were observed that might account for this response.

Questions

1. Define cuttage.
2. Give some advantages of propagating plants by cuttage.
3. Why are "own-rooted" plants sometimes not desirable?
4. Name the different classes of cuttings.
5. Give some disadvantages in the use of root cuttings.
6. How does a new plant form from a leaf cutting?
7. Distinguish between herbaceous and softwood cuttings.
8. What are the advantages of callusing cuttings?

9. What plants may be grown from cuttings started in the field?
10. Give various media used in rooting cuttings.
11. How are cuttings treated with growth substances?
12. Should leaves be removed from cuttings? Why?
13. What type of cut should be made at the base of the cutting?
14. Where should the basal cut be made?
15. In what way may stored food influence rooting?
16. What is etiolation and how may it be applied in rooting of cuttings?

Suggested References

1. Chadwick, L. C.: Studies in Plant Propagation, *N. Y. (Cornell) Agr. Exp. Sta. Bull.* 51, 1933.
2. Curtis, O. F.: Stimulation of Root Growth in Cuttings by Treatment with Chemical Compounds, *N. Y. (Cornell) Agr. Exp. Sta. Mem.* 14, 1918.
3. Halma, F. F.: The Propagation of Citrus by Cuttings, *Hilgardia* **6**: 131–157, 1931.
4. Hitchcock, A. E., and P. W. Zimmerman: Relation of Rooting Response to Age of Tissue at the Base of Greenwood Cuttings, *Contrib. Boyce Thompson Inst.*, **4**: 85–98, 1932.
5. ——— and ———: Effect of Growth Substances on the Rooting Response of Cuttings, *Contrib. Boyce Thompson Inst.*, **8**: 63–79, 1936.
6. Johnson, Stanley: The Propagation of the Highbush Blueberry, *Mich. Agr. Exp. Sta. Sp. Bull.* 202, 1930.
7. Laurie, Alex: The Use of Peat in the Greenhouse, *Mich. Agr. Exp. Sta. Sp. Bull.* 194, 1930.
8. Pope, Willis T.: Propagation of Plants by Cuttings in Hawaii, *Hawaii Agr. Exp. Sta. Circ.* 9, 1934.
9. Stoutemyer, V. T.: Stimulation of Hardwood Cuttings by Means of Synthetic Growth Substances, *Amer. Nurseryman*, LXVIII, No. 9, 3–5, 1938.
10. Tukey, H. B., and Karl D. Brase: The Propagation of Multiflora Rootstocks for Roses by Softwood Cuttings, *N. Y. (Geneva) Agr. Exp. Sta. Bull.* 598, 1931.
11. Winkler, A. J.: Some Factors Influencing the Rooting of Vine Cuttings, *Hilgardia*, **2**: 330–349, 1927.
12. Van der Lek, H. A. A.: Root Development in Woody Cuttings, *Meded. Landbouwhoogesch*, (Wageningen) deel 28, Verhandel 1, 230 P. (English Summary pp. 211–230), 1924.

CHAPTER X

GRAFTAGE

Graftage is the art of inserting a part of one plant into another plant in such a way that the two will unite and continue their growth. It differs from cuttage, layerage, and bulb propagation in that the plant part expected to produce the top of the new plant is deprived of its own root system and unites with another plant that supplies this part.

FIG. 91.—An oriental pear tree top-worked with the Garber variety. Seven buds were inserted in the framework of a 3-year-old rootstock early in April, and the picture shows them as they began growth the following March. The points marked with an *X* indicate places where the T buds were inserted.

The art of graftage is not new. Contrary to popular opinion, it is no recent innovation in the arts of plant craft. Pliny, writing before the birth of Christ, recognized graftage as a horticultural practice, and it is known that it was practiced before his time. Columella, who died shortly after the birth of Christ, mentioned certain kinds of graftage, particularly the bark graft, cleft graft, and patch bud, which he said had been practiced by the ancients. It is a significant fact that at those early periods in agricultural history the unreliability of seeds and the importance of graftage were appreciated in the reproduction of varieties. At various times, including the present era, many methods known in ancient time have actually been rediscovered by workers who were not familiar with their previous use.

The field of graftage includes *scion grafting* and *bud grafting*, commonly referred to as *grafting* and *budding*. The two, however, are so different that a discussion of each will be reserved for separate chapters. Some of the operations and terms that

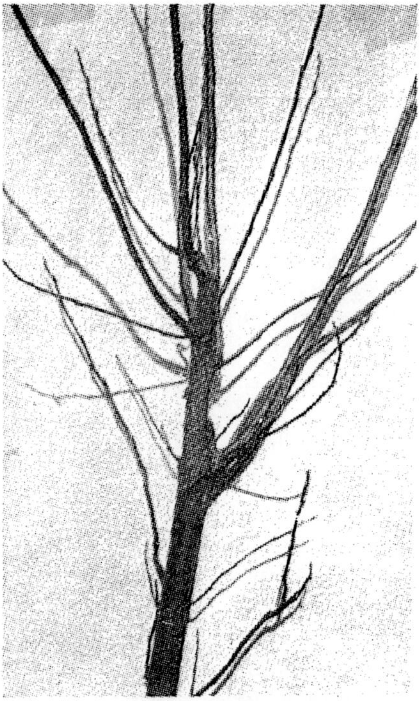

FIG. 92.—Showing the development of one-year-old shoots from a stub of a dehorned tree.

are common in all types of graftage will be considered in the following sections.

Top-working.—The series of operations whereby the top of a plant is replaced with a top of a different variety is known as *top-working*. In some cases a large part of the old top is cut away and a new one started; in others the new top is started, after which the old one is cut away by degrees until the top consists largely of a different variety. Trees may be top-worked successfully by either budding or grafting or by a combination of the two. The process may be completed within one season or it may extend over several years, depending upon the size and conformation of trees to be top-worked. In reality, budding or grafting of small nursery trees is top-working; the term, however, is generally used with regard to changing the tops of larger trees.

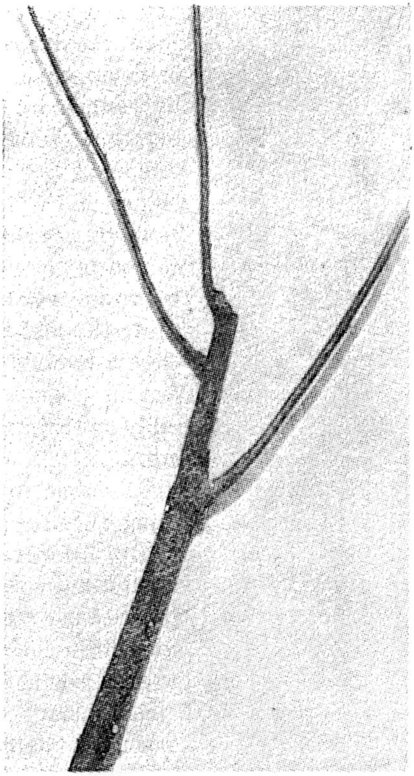

FIG. 93.—The same stub shown in Fig. 92; the shoots have been thinned for budding.

Dehorning.—The practice of cutting the main limbs and trunk of a tree back to stubs is known as *dehorning*. The extent to which trees can be safely cut back varies with the species; some can be cut back much more severely than others. In practice, trees are cut back so that the stubs that remain range in length from 1 to 4 feet and in diameter from 1 to 6 inches. When a tree is dehorned the limbs should, if possible, be cut at points that will result in the new top having a symmetrical shape. It is not advisable to cut limbs at different heights so that the new growth of some will obstruct sunlight and create shade for others. In order to facilitate healing of the wound a limb to be dehorned should be cut at a point where a side limb or a lateral bud occurs on the upper side. This virtually assures growth from very near the terminal end of the stub, and this encourages overwalling of the wound. If no shoot grows within 1 inch of the end of the stub, it usually becomes advisable to recut it during the first year at a point where a lateral limb has developed in the meantime, preferably on the upper side of the stub.

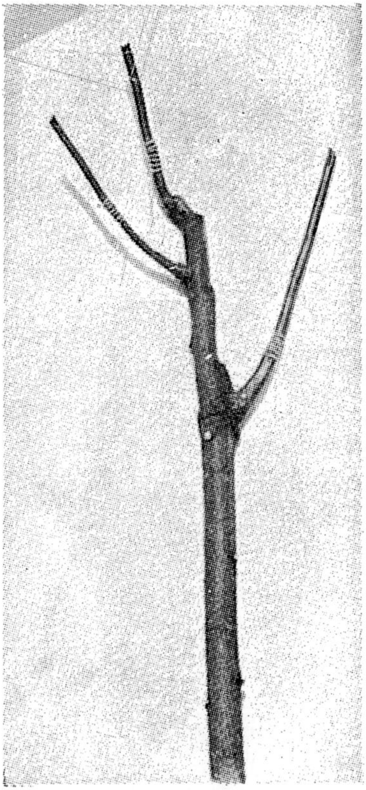

FIG. 94.—Showing the placing of buds on the shoots shown in Fig. 93.

The season for dehorning is during the dormant period, shortly before growth is resumed in the springtime. The practice has considerable application in renovation pruning and is used in many cases as a preliminary step in the top-working of large trees.

Forcing.—Any treatment that encourages and hastens growth of a bud or graft is referred to as *forcing*. Essentially, it consists of creating for the bud or graft a terminal position from a physiological standpoint. In practice it may be accomplished by (1) cutting the stock off above the bud, by (2) girdling above the bud, by (3) allowing the tying material to bind above the bud, and by (4) bending or breaking the stock above the bud without severing it. Buds or grafts that are exposed to sunlight are more easily forced into growth than those that are shaded.

The size of the stock and growth habits of a plant should determine the extent to which forcing is practiced. If forced

FIG. 95.—Peach tree of appropriate size for top-working.

FIG. 96.—Same tree as shown in Fig. 95, dehorned in late winter. New shoots

too much, there is the likelihood of the graft or bud becoming top-heavy and breaking during a wind, rain, or ice storm. This is especially likely to happen if the stock is large and inflexible; the pressure of a force in such a case is applied at the point of union where vascular tissue connections may not be strong enough to withstand the strain. Terminal grafts, because of their position, seldom require forcing; their growth, however,

FIG. 97.—Pecan tree of suitable size to be top-worked.

can be influenced by the extent to which native shoots on the stock below the graft are cut back.

Stocks.—The term *stock* has several different meanings when used in connection with propagation. *Stock plants* are those that are grown, frequently in a greenhouse, as a source of propagation material, such as cuttings, layers, and sometimes buds or grafts. *Nursery stock* means plants that are grown by nurserymen, usually for sale. *Lining-out stock* refers to plants that are of suitable size to be replanted or lined out in the nursery row and allowed to grow for one or more seasons, during which time they become large enough for a designated use. A *rootstock* is the plant that supplies the root system, and in many cases a part of the trunk and framework, for a budded or grafted tree; rootstocks are commonly grown from seeds, cuttings, and, rarely, layers. The term *stock*, as frequently used, is synonymous with rootstock. The term *rootstock* is also used to describe underground stems, such as rhizomes, and this causes some confusion in interpretation of horticultural literature. In propagation work, plants that are grown from seed are known as *seedlings*, until they are budded or grafted, after which they become *seedling rootstocks*. Those plants that are propagated by asexual methods, usually cuttage and occasionally layerage from a single plant or variety, are known as *clonal rootstocks* after they have been successfully budded or grafted. When grafts are placed on unrooted cuttings, the latter are called *cutting stocks*. An *intermediate stock* is one that separates and connects a rootstock and a budded or grafted top, yet is different from both; it is also known as a *splice*. The term *body stock* is used to indicate that the trunk and framework of a tree are different from the budded or grafted top. The body stock may be the same as the root system, but in many cases it is of a different kind, as a consequence of double-working.

Fig. 98.—Pecan tree dehorned to cause growth of new sprouts suitable for budding (see Fig. 97).

Matrix.—The matrix is a place on the rootstock that is prepared for the insertion of a bud or graft.

Scions.—The limbs that are cut from any plant to be used in graftage are known as *scions*. Those which are to be used for grafting are known as *grafts*, or *graftwood;* and the ones that are to be used as a source of *buds* for budding are called *budwood.*

Scions reproduce the kind of tree or plant from which they are taken and hence are obtained from the variety to be propagated. Healthy parent plants should be selected, in order to prevent the spreading of disease in propagation.

The success of budding and grafting by the different methods depends, among other things, upon the use of the appropriate kind or type of scions, and also upon methods of handling them

FIG. 99.—A pear tree of a suitable size for top-working.

from the time they are cut from the parent tree until they are finally used. The time that intervenes may be a few hours or several months.

Graftwood.—Scions for grafting are usually obtained from one-year-old wood; sometimes older wood is used. They should be straight and smooth and have normal, plump buds and few or no side branches. The size range for graftwood may vary considerably for different methods.

Scions for grafting should be thoroughly dormant at the time they are used. They should be secured before the plant from which they are to be taken shows any signs of growth. Those cut in midwinter tend to remain dormant longer after they are inserted into a stock than those that are cut at a later date; scions that make premature top growth before union is established usually wither and die within a few days. In practice it is customary to cut scions any time from midwinter until 2 or 3

FIG. 100.—The same pear tree shown in Fig. 99, with top cut back and inlay grafts inserted.

weeks before the parent tree begins growth in the spring. These grafts are frequently used for grafting as they are secured from the parent tree, with no more than a day or two intervening.

Scions that are to be used relatively late in the grafting season may be cut and held in cold storage in order to keep them dormant. At a temperature of from 32 to 36°F. graftwood can be stored successfully for 4 months or longer, though there is seldom any occasion for storing it this long. Prior to storage the scions should be packed in moist insulating material. Sphagnum moss and coarse sawdust are commonly used for this purpose; they are light and easy to handle, absorb moisture readily and retain it well. The packing material should be moist, though not so wet as to cause poor aeration. Tests with storing pecan scions show that from 3 to 5 pounds of water for each pound of sphagnum moss is sufficient. It is desirable to

FIG. 101.—Showing (1) proper and (2) improper way to cut limbs in dehorning.

add the water and blend it uniformly with the insulating material before packing it around the scions. Alternate layers of the insulation and scions are placed in a box of convenient size for storage. It is important that the packing be pressed firmly into all corners and around the edges of the box. Heavy paper folded over the box will restrict evaporation and delay the time when additional moisture will be needed. If packed properly and held at a low temperature, scions remain thoroughly dormant and may be used directly out of storage throughout the springtime.

Scions should be properly labeled with the name of the variety at the time they are cut; otherwise, different varieties are likely to become mixed. Failure to preserve the identity of scions is serious, because mixing of varieties is oftentimes not discovered until the trees come into bearing.

Budwood.—Several different types or kinds of scions are used as a source of buds for budding. The classification of them is

FIG. 102.—Citrus-seedling rootstocks. Left to right, Rusk citrange, Cleopatra mandarin, rough lemon, sour orange, sweet orange, Tresca pomelo.

based largely on the age of the wood and its condition of growth or dormancy at the time of cutting or use.

1. *Current-season Scions.*—Buds that are taken from limbs that are in their first season of growth are suitable for some methods of budding. When so used, they are known as *current-season buds.* The habits of growth of plants determine in a large measure how early in the growing season such buds may be used. On some plants they mature sufficiently for use within 6 weeks or 2 months after growth begins in the springtime; on others they do not mature until much later. In the South it is cus-

tomary to use peach buds of the current season's growth as early as June; pecan buds are not sufficiently mature until late in July.

Current-season budwood may be "ripened" by cutting, at a point ½ or ¾ inch out from the base, leaf petioles that subtend each bud. This is done before the scions are cut from the

FIG. 103.—Showing method of packing budwood to be placed in cold storage; another layer of moss is placed over the budwood.

parent tree; it causes the petiole to fall off within 6 or 8 days, and a corky covering forms over the petiole scar. If the leaf petioles are cut too far in advance of the time for cutting the scions, buds in corresponding axils are likely to force into growth and not be suitable for budding purposes. In ripening budwood the petioles should be cut from only the basal part of each shoot, so that perhaps only two-thirds of the leaves are removed. The buds near the terminal are usually too immature for use, and there is no object in removing the leaves from that portion.

All leaves should be removed from current-season scions as soon as they are cut from the tree in order to restrict transpiration. Buds from them should be used as soon as possible after they are obtained. If they cannot be used immediately they may be kept for a week or longer, packed in moist material and stored at a temperature of from 34 to 38°F.

FIG. 104.—(1) Dormant apple scions; (2) current-season budwood of rose before and after the leaves are removed.

2. *Previous-season Scions.*—For early spring budding it is necessary to use budwood that grew during a previous season. Limbs that have made normally vigorous growth during the preceding season are preferred. The basal and midportions of one-year-old wood furnish the best buds; those buds on the angular, small, or immature wood of the terminal portion are not usually satisfactory. Buds may be used from two- or three-year-old wood of some plants, but they are seldom entirely

satisfactory. There are two different methods of using previous-season wood:

a. Fresh scions are cut from previous-season growth of the parent tree as needed throughout the season for budding. Budwood of some plants remains in acceptable condition for use throughout the season; on other plants the best buds are forced into growth in the early springtime, and buds that are satisfactory for use after that time are scarce.

b. Storage budwood is used extensively in the propagation of certain plants. Two general practices are followed in using it. According to one of these it is cut during the dormant season and packed and stored under conditions similar to those prescribed for graftwood, and at the same temperature. For certain methods of budding it may be used directly out of cold storage in a dormant condition. If, on the contrary, it is to be used in one of the methods that requires that the bark separate from the wood, the budwood must be subjected to conditions upon removal from storage that will cause the cambium layer to become active. This treatment, known as *seasoning*, is accomplished by providing ample moisture to prevent drying out and a temperature of from 78 to 85°F. It is considered that budwood is seasoned when the bark and buds can be peeled readily from the wood. The number of days required for budwood to become seasoned varies commonly from 3 to 10 days. Seasoned budwood may be used immediately; or it may be returned to cold storage and held for as long as 1 month. It should be noted here that dormant budwood may be kept for several months, but seasoned budwood for a limited period only.

By another practice previous-season budwood is cut, packed, and stored as soon as the bark will slip but before the buds have made any perceptible growth. Budwood of this kind will keep satisfactorily for about 1 month at a temperature of 32 to 36°F. The cambium layer remains active in cold storage and the wood can be used as soon as it is removed.

The Bud or Graft Union.—The primary objective of graftage is to secure the union of the stock and scion. The success or failure of the bud or graft depends upon this union, which normally takes place as a result of the formation and commingling of callus by both the stock and scion. This callus is produced by the cambium layer; it may be seen on cut ends of cuttings and often on whip grafts as a spongy mass of white tissue. When the scion is placed in the matrix on the stock a definite effort is made to have the two cambium layers coincide, or match. If, as a result of poor technique, the scion and stock fit poorly, a greater amount of callus will be necessary, a longer time will be required, and the chances of ultimate union are

FIG. 105.—A live oak top growing on a post oak rootstock.

less than if a better fit were obtained. A firm pressure is always necessary to produce a graft union, because the callus formed by the stock and that formed by the scion tend to spread or separate the two component parts. For this reason they must be held in place by tying if necessary until a union is formed. Many bark grafts especially are lost because the strings are removed too soon. Union of a graft and stock takes place only in tissue that forms after the graft is made; the original woody portion of a scion never unites with a stock. The proper placing of a scion affects not only the probability of its growing but also the strength of the resulting union.

Callus formation is influenced by the vigor and species of a plant, by the presence of inhibitors, such as gums, resins, and

FIG. 106.—Upper part of this tree is live oak, top-worked onto a post oak.

latex, and by the environmental factors of temperature and humidity.

Limits of Graftage.—In commercial practice, graftage is limited to those plants that have a continuous cambium layer between the xylem and phloem. In addition, plants must have some degree of botanical relationship in order to unite.

1. *Grafts between Varieties.*—As a general rule a horticultural variety may be budded or grafted upon any other variety of the

same species. The Elberta peach will unite with the J. H. Hale peach, and the Delicious apple with the Jonathan, for example.

2. *Grafts between Species.*—It is generally true that species of the same genus may be cross-grafted, or budded, one upon the other. Peach will grow on plum, almond, cherry, or apricot, for example, forming good unions with some and poor unions with others; pecan will unite with the bitter pecan and with hickory; California privet and Amur privet are used as rootstocks

FIG. 107.—Pecan on seedling rootstock. Note overgrowth below union.

for waxleaf ligustrum. All these are examples of plants of species that will unite with other species of the same genus, some forming successful and others unsuccessful combinations.

3. *Grafts between Genera.*—Plants that belong to different genera of the same family cannot be intergrafted with any certainty of success. There are examples of some that will unite, and many others that apparently will not. The apple unites with the hawthorne, and chestnut with white oak. The pecan is reported as having been grown on black walnut for a

part of one season. In general such wide crosses do not form satisfactory combinations. Chestnut on oak seldom lives longer than one or two seasons. Apple tops may live for several years on hawthorne rootstocks, but they seldom grow or fruit in a normal manner. It is not possible to graft a stone fruit, as the peach, onto a pome fruit, as the apple or pear, by methods ordinarily used in propagation of these plants.

Stock and Scion Effects.—The term *stion* is used to designate relationship between a stock and scion; and *stionic effects* are the reciprocal influences of stock and scion. The fact that a

FIG. 108.—Delmas pecan top-worked on hickory rootstock. Note marked overgrowth. (*Courtesy J. F. Rosborough, Tex. Ext. Ser.*)

stock and scion will unite is no assurance that the union will be strong or enduring. Oftentimes the response of a given top on a root system is not predictable on the basis of botanical relationship. Unions between certain plants are successful; those of others may succeed for a time but ultimately fail. The immediate cause of failure may be due to either the rootstock or the scion top. Failure of either is quite soon reflected in the functioning of the other. The basis of the success or failure of a stionic union is either anatomical or physiological in a final analysis. *Anatomical weaknesses* are reflected in the characteristic overgrowths of stock in some cases, and scions in others; and also in the alternating unions and "breaks" that occur in the cambial region at the point where the two meet. *Physiological uncongeniality* is due to inability of the stock or scion to supply adequately the other component with the necessary materials for normal functioning with regard either to amount or quality. Whatever the relationship with reference to stionic congeniality may be, the effect is to influence the vigor

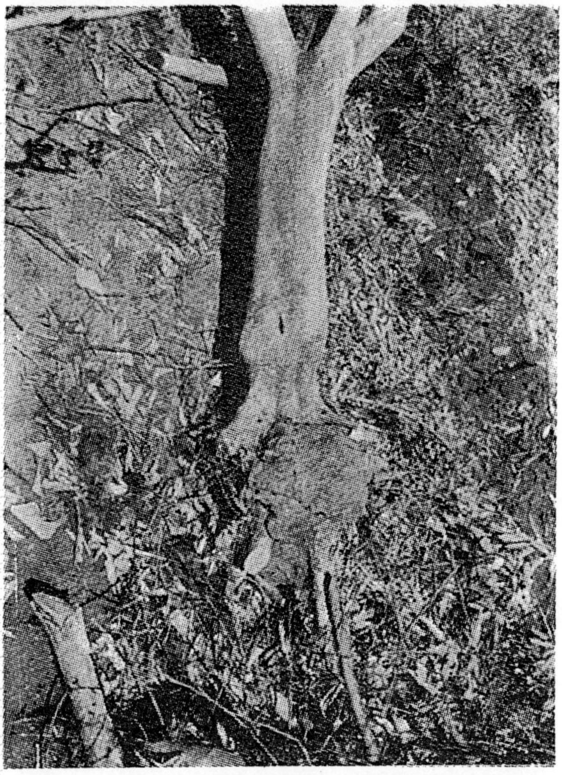

FIG. 109.—Grapefruit on sour orange rootstock. Note abnormal overgrowth of top above union.

of the plant. Vigor in turn influences indirectly the quality and size of fruit; yields; earliness of bearing; earliness of maturity; hardiness to heat, cold, and drought; and stature, some being dwarfed by a certain rootstock and invigorated by another. Rootstocks vary in soil adaptation, tolerance of insects and diseases, and in other respects; and they should be selected for a variety with the same careful consideration given cultural practices.

Double-working.—Double-working is a practice in which two successive budding or grafting operations are performed on the same plant. After one scion has been placed on a stock a second scion is grafted or budded into the first one. This procedure results in a tree composed of three different kinds or varieties of

FIG. 110.—Kieffer pear on Japanese pear rootstock. Budded shoot on the left center had begun growth in the spring, while the stock was apparently still dormant.

wood; the first scion inserted becomes a splice or intermediate stock between the root system of the original stock and the new top of the double-worked tree. The first scion is generally allowed to grow a year before it in turn is top-worked to another variety. However, buds of the desired variety may be set into the limb that is to serve as the splice before it is severed from the parent plant. The bud remains dormant until the splice is grafted onto the original stock, when it begins growth, resulting in a double-worked tree. The two grafting operations may be performed at the same time with some plants, though considerable skill is required to insure successful unions. There are several applications of this method, especially in the production of fruit trees.

1. Uncongenial stocks and scions may be united by means of a splice that is congenial to both. Certain pear varieties, such as Bosc and Winter Nelis, do not form good unions with quince, and if it is desirable to grow these varieties on quince rootstocks they are double-worked by using a splice of Bartlett, Angouleme, or some other variety that does make a good union with the quince stock and the pear top.

2. Resistance to specific troubles may also be secured by double working. Certain varieties of apple may be used as a splice to give resistance to collar rot. Injuries to the crown of the tree, due to freezing, may be overcome or lessened in the same manner.

3. Top-working of orchard trees that have been grafted originally results incidentally in double-working. In such cases the intermediate stock or stem continues to exert a certain influence on the root system of the double-worked tree and may also affect the new top as well. If a tree of a certain variety is being top-worked because of some inherent weakness, there is a definite possibility that such a practice will not overcome this weakness.

The performance of a grafted or budded tree is determined by the reciprocal influences of both the stock and the top. The possibility of uncongeniality is increased where there are two graft unions instead of one, and three varieties of wood instead of two. Varieties should not be double-worked without regard to the intercongeniality of all three components.

Graft Hybrids or Chimeras.—The term *graft hybrid* is a very misleading one; it fosters the rather common conception that hybrids are the normal result of grafting. On the contrary, one of the primary objects of graftage is to maintain a variety true to type.

A *graft hybrid* is a stem, branch, or plant originating from an adventitious bud at the graft union. Not all such buds produce graft hybrids—only those do so which contain tissue of both stock and scion. The resulting shoot and its fruit will be hybrids only from the standpoint of morphology and not from the genetic viewpoint. The graft hybrid has been designated by some writers as a special form of graft symbiosis.

Chimera is a term commonly applied to graft hybrids and to similar phenomena occurring naturally. In one case where the two masses of tissue meet, the plant and fruit will consist of sectors of tissue from both stock and scion; this type is known as a *sectorial* chimera. In another case, the tissues of stock or scion overgrow those of the other, to form a *periclinal chimera.* A third type, the *hyperchimera,* is produced by a vegetative cone that is a mosaic of unlike cells.

Objects of Graftage.—Graftage may be employed to increase the usefulness of plants in a number of different ways.

1. It is a way of preserving and perpetuating some varieties that cannot be reproduced easily by other vegetative methods. When a relatively simple method like cuttage is not effective, graftage is often used successfully. Standard varieties of peaches, apples, walnuts, pecans, and many other fruits are not easily propagated by asexual methods other than graftage.

2. Graftage makes it possible to change trees of poor varieties to varieties that are considered to be more desirable. A tree might be undesirable by being nonproductive, a shy bearer, or an irregular bearer; it might have undesirable habits of growth, an untimely blossoming period, or an untimely ripening period of fruit. Trees of considerable size when top-worked will resume production of fruit and produce more heavily in a shorter period of time than nursery trees.

3. Adaptation to unfavorable environment may be accomplished by graftage. In many cases, stocks may be found that are especially resistant to some soil condition, as heavy soil, poor drainage, acidity, or alkalinity; to some insect or similar pest living in the soil, as grape root louse, woolly aphis, or nematode; or to some particular disease, as foot rot or gummosis of citrus. In the case of unfavorable soil conditions, it is often advisable to abandon the location in favor of one better adapted; but in the case of insects and diseases, resistant stocks frequently give entirely satisfactory results.

4. Graftage is a means of hastening new varieties into bearing. Scions secured from young trees and caused to grow on mature trees reach bearing stage sooner than if left on their own roots.

Several new varieties may be tested out on one large tree and thus space be conserved. This method may be used to advantage in fruit-breeding work.

5. The development of seed and fruit by dioecious plants may be encouraged by grafting or budding in scions of the lacking sex. Varieties that are incompatible or dichogamous may be rendered more fruitful by grafting in scions of varieties that are known to be good pollinators.

6. Graftage is used to encourage healing of tree wounds caused by implements, disease, injurious temperature, rabbits, mice, or other rodents.

7. The novelty of growing several different kinds of fruit or flowers on one plant is made possible by graftage. A pecan tree is on record that has thirteen different varieties of pecans and one hickory nut variety growing on it.

Questions

1. How long has graftage been practiced?
2. What is top-working? What determines the length of time required to top-work a tree?
3. What are the objects of dehorning? What rules should be observed in dehorning a tree?
4. How is forcing accomplished?
5. What is the meaning of: Stock plant? Nursery stock? Lining-out Stock? Rootstock? Seedling rootstock? Clonal rootstock? Matrix?
6. Tell how to select, prepare, and store graftwood.
7. What are the different classes or types of budwood? How long can each be stored? What is "seasoning"?
8. Explain how union takes place between a stock and a scion. What factors influence the formation of callus?
9. What factors determine whether plants can be successfully grafted?
10. In what ways does the rootstock influence the grafted top of a tree? What are the causes of unsuccessful graft unions?
11. What are the various objects of graftage?

Suggested References

1. BONNETT, L. O.: Phylloxera Resistant Vineyards, *Calif. Agr. Exp. Sta. Circ.* 288, 1925.
2. BRADFORD, F. C., and B. G. SITTON: Defective Graft Unions in the Apple and the Pear, *Mich. Agr. Exp. Sta. Tech. Bull.* 99, 1929.
3. BRISON, FRED R.: The Storage and Seasoning of Pecan Budwood, *Tex. Agr. Exp. Sta. Bull.* 478, 1933.
4. CHILDS, LEROY, and GORDON G. BROWN: A Study of Tree Stocks in Relation to Winter Injury and Its Prevention, *Ore. Agr. Exp. Sta. Circ.* 103, 1931.

5. Jacob, H. E.: Propagation of Grape Vines, *Calif. Agr. Ext. Serv. Circ.* 101, 1936.

6. Roberts, R. H.: Some Stock and Scion Observations on Apple Trees, *Wis. Agr. Exp. Sta. Research Bull.* 94, 1929.

7. ———: Top and Double-Working Apple Trees, *Wis. Agr. Exp. Sta. Bull.* 432, 1936.

8. Schuster, C. E.: Grafting and Budding, *Ore. Agr. Ext. Serv. Bull.* 491, 1936.

9. Stoutemyer, V. T.: Regeneration in Various Types of Apple Wood, *Iowa Agr. Exp. Sta. Research Bull.* 220, 1937.

10. Talbert, T. J.: Top and Double Working, and Bridge Grafting of Fruit Trees, *Mo. Agr. Exp. Sta. Circ.* 196, 1938.

11. Tukey, H. B., and Karl D. Brase: Influence of the Cion and of an Intermediary Stem-Piece upon the Character and Development of Roots of Young Apple Trees, *N. Y. (Geneva) Agr. Exp. Sta. Tech. Bull.* 218, 1933.

CHAPTER XI

GRAFTING WAXES, MATERIALS, AND TOOLS

The various practices of budding and grafting, to be discussed in succeeding sections, entail the use of a wide diversity of materials and tools. It is appropriate that attention be directed to these before the operations in which they are used are considered.

Waxing Materials.—The most satisfactory materials for covering graft and bud unions are waxes, different kinds of which may be applied in melted condition, or spread on as a semisolid or paste. The value of these materials is determined by several qualities, such as ease of application, adherence to the plant without cracking or melting, and freedom from compounds injurious to plant tissue.

1. *Hand Wax.*—This is the standard wax that has been used generally for many years. The formula includes rosin, 4 parts; beeswax, 2 parts; and tallow, 1 part. The tallow is melted first and then the beeswax; the rosin is added last. When all the materials are melted the mixture is poured into cold water. As soon as it cools sufficiently it is removed and worked with greased hands until smooth and light colored. It is then molded into balls and wrapped in waxed paper. In grafting operations this wax is worked with the hands into thin strips and applied without heating.

2. *Alcoholic Wax.*—Alcoholic wax, made according to the following formula, has long been a popular material.

Pine rosin	2 pounds
Beeswax	1 pound
Alcohol	about 180 cubic centimeters

The rosin and beeswax should be melted together over a slow fire but not allowed to boil. When the mixture is completely melted it should be removed from the fire and allowed to cool. It should be stirred constantly to prevent uneven cooling and adhering to the sides of the container. The cooling process may be hastened by pouring back and forth from one bucket into another, or by partially submerging the container in cold water. When it has cooled to such a degree that it feels only warm, the alcohol is added, about a tablespoonful at a time, with constant stirring. This should be continued until the mixture is granular in texture and yellow in color. Small amounts taken from the mass at intervals cool quickly and serve to indicate when suffi-

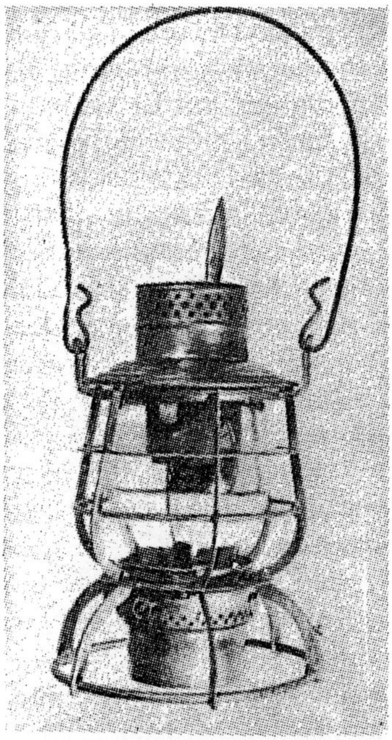

FIG. 111.—Good type of paraffin melter.

cient alcohol has been added. For cool weather, more alcohol is used, and, for warm weather, less. Beating should be avoided, as it tends to incorporate air bubbles in the wax, which decreases its adhesiveness. Pure grain alcohol, wood alcohol, or denatured alcohol may be used.

3. *Hard Wax.*—A number of different waxes are used, with slight variations in ingredients, which must be melted and applied with a brush. One of the best of these, which is tough and will not crack, is made of 4 pounds of rosin, 1 pound of beeswax, 1 pint of linseed oil or 1 pound of tallow, and 1 ounce of

lampblack. The beeswax and rosin are melted first, and the linseed oil then added. The mixture is removed from the fire, and the lampblack is stirred in slowly.

4. *Paraffin.*—Within recent years, melted paraffin has partially replaced grafting wax. The range of temperatures at which the different commercial grades melt is from 40 to 50°C. A paraffin wax with a low melting point is preferred, since its use involves less danger of heat injury to the bud. For field or nursery work, the paraffin may be used from a thermos bottle that has been previously filled with melted paraffin. Specially devised paraffin heaters are on the market. When these are used, paraffin in the open-top containers is kept melted by the flame of an alcohol lamp. Precaution should be exercised to prevent it from becoming so hot as to injure the buds to which it is applied. It may be used in connection with practically all the methods of budding and grafting, being especially convenient for use with skin buds, which are small. A brush is taken from the melted paraffin and drawn briskly across the bud from each side, the result being a thin film over the buds and all cut surfaces.

Paraffin is likely to check and crack in cool weather, and in very warm weather it may cause some injury to the plant.

5. *Paraffin Wax.*—Some of the objections to the use of paraffin are overcome in a wax made of rosin, 1 pound; raw linseed oil, 3 fluid ounces; paraffin, 5 pounds. This mixture is useful as a brush wax.

Waxed Tying Materials.—A suitable mixture for making waxed tying materials can be prepared by the following formula:

Rosin	4 pounds
Beeswax	2 pounds
Rendered tallow	1 pound
Paraffin	½ pound
Linseed oil	1 pint

These ingredients are placed in a container and heated until all are melted. The linseed oil and tallow cause the mixture to be soft and sticky; smaller amounts of these than indicated may be used for materials to be used in warm weather.

1. *Waxed string* is made by placing small balls of cotton twine, usually four-ply, in the hot mixture and allowing them to remain until they become thoroughly saturated. Fifteen minutes is usually sufficient time for this, after which the balls are removed

and allowed to drain and cool for future use. A few inches of string should be unwound from each ball before it is placed in the mixture, in order that the end of the string may be easily located after it is taken from the wax. Waxed string is convenient for wrapping grafts, as the wax will hold the string in place and no tying is necessary.

2. *Waxed patches* are made by treating domestic cloth with the hot mixture of rosin, beeswax, tallow, paraffin, and linseed oil. Strips of the domestic are dipped in the hot solution and withdrawn from it between two boards to cause the excess wax to drain off. When strips have cooled, they are folded back and forth in such a way as to form 1½- or 2-inch squares or patches. A hole is made in the center, through which the bud extends when the patch is used. It should be adjusted carefully over the bud, pressed well about it and the stock so as to exclude air and water, and tied with several wrappings of cotton twine. Waxed patches are used extensively in chip, patch, and ring budding.

3. *Waxed tape* is prepared by tearing domestic cloth into bands the width of which is the desired length of the waxed strip. These bands are rolled lengthwise into a large roll and dipped in the hot solution of the formula given above. The roll should be allowed to remain several minutes to insure thorough penetration, after which it is removed and allowed to drain and cool. Waxed strips from ⅓ to ½ inch in width are torn crosswise from the roll as needed. Size of stock upon which they are to be used should determine the length of the strips. A desirable length is from 12 to 15 inches, which means that the roll of domestic, when prepared, should be of that width. Waxed tape is wrapped in such a way as to exclude air and water and to hold buds securely in place, no additional tying being necessary. A good grade of domestic cloth should be used, as strips cannot be torn satisfactorily from a poor grade.

Tying Materials.—Nursery practice throughout the country has resulted in the utilization of many materials for tying buds and grafts. In many cases the material most readily available has been used, without due regard to its actual value.

1. *Cotton twine* is a popular tying material for buds and for grafts on small stock. Injuries caused by the string cutting into the tissues, causing "binding" and "strangulation," are

less frequent if a heavy grade of twine is used. Ten- or twelve-ply twine is especially suitable for tying grafts that are inserted in large stocks. Such grafts require more support than is offered by smaller twine.

2. *Raffia*, the epidermis of a Madagascar palm, is used extensively in tying buds. It is soft, strong, and durable and can be torn in strips of convenient width. Used while moist, it is easy to handle and does not have so strong a tendency to roll as when dry. Incidentally, raffia is a convenient material to have around the propagating room or greenhouse. It is used in tying flowers, celery, bundles of asparagus and in staking vegetables and other plants.

3. *Corn shucks* are commonly used by many nurserymen of the South for tying T buds. They are torn into strips ¼ or ⅓ inch wide, moistened to facilitate tying, and tied into small bundles with ends exposed in a way that will permit individual strips to be conveniently drawn from the bundle as needed.

4. *Rubber budding strips* that are especially adapted for tying buds or even grafts have been produced in recent years. These strips are made of gray or red rubber, in widths of 3/32 to ⅜ inch and in lengths of 3½ to 7 inches. The gray strips deteriorate more quickly than red strips of the same weight. Advantages that are pointed out for these strips are faster application, even pressure, greater protection, and the fact that they are self-releasing through deterioration. The use of the rubber strips has become quite general, especially for nursery work. They are recommended also for bench grafting evergreens and roses.

5. *Nursery tape*, for the purpose of controlling root knots, is used with success on whip grafts. The material is available in widths of ¼ and ½ inch in rolls of 60 yards. A dispenser with a cutting edge enables the operator to use it conveniently. Many nurserymen regard the use of this tape as a most important contribution to propagation work.

Knives.—The knife is a tool of universal application in all horticultural work. Individuals have personal preferences as to shape and size of blade for budding, grafting, or pruning, but all are agreed that they should be made of good metal that will take and hold an edge.

1. *Budding knives* are made with handles of various shapes and materials, but the general shape of the blade is uniform. It

should have a straight edge, with the end rounded from front to back. The straight part of the blade is used for cutting the buds, and the curved portion for making the incisions in the stocks. Some budding blades have a blunt projection at the end, on the back of the blade, which is designed for use in lifting the bark in T budding; other knives have a bone, ivory, or celluloid projection, at the end opposite the blade, to be used for the same purpose. Ordinarily, when the bark on the stock slips freely, as it should for budding work, there is no need for such a device; budders seldom lift the bark before inserting the bud.

In large-scale nursery operations the knife most commonly used is made of a good steel blade mounted rigidly in a plain wooden handle.

For grafting or chip budding a knife should have a budding blade of the same shape but more substantially constructed than the one described above. It is even more important that knives for such purposes be made of the finest steel; and they should be kept razor sharp in order that the different cuts can be made with precision.

2. *Grafting knives* are made with a blade that is straight, or very slightly curved inward, along its entire edge. The blade is heavier than that of a budding knife. In some of the best of these knives the blade is beveled from one side only. As in the case of budding knives, these knives are made with stationary blades or with folding blades.

3. *Pruning knives* are quite different in appearance from others, in that the blade has a very pronounced hook shape. The pruning hook is one of the oldest of horticultural tools, but its use in this country is rather restricted.

4. *Patch budders* are of two general types. One consists of four blades mounted on a short metal handle. It is used as a stamp and is satisfactory if the stock and budwood are of about the same size.

In the other type two blades are mounted on a wooden handle so as to be parallel and rigid. Two ordinary budding knives can be made into such a tool by riveting them onto a wooden handle so as to make the blades parallel and ¾ or ⅞ inch apart. Two safety razor blades constitute the cutting edges in a tool of slightly different design. The materials needed for making it are two 1½-inch stove bolts, two safety razor blades, and a piece of seasoned hardwood timber 8 inches long and 1¼ inches wide and 1 inch thick. Two longitudinal cuts, parallel and ⅞ inch apart, should be made with a fine-toothed saw perpendicular to the 1¼-inch surface. They should extend down 5 inches from one end. Two holes should be drilled, 1 inch and 3 inches, respectively, from the same end, through which the 3/16-inch stove bolts are inserted. These holes are drilled through median lines of the 1-inch surfaces of the timber. The surface between the bolts on one of the 1¼-inch sides should be

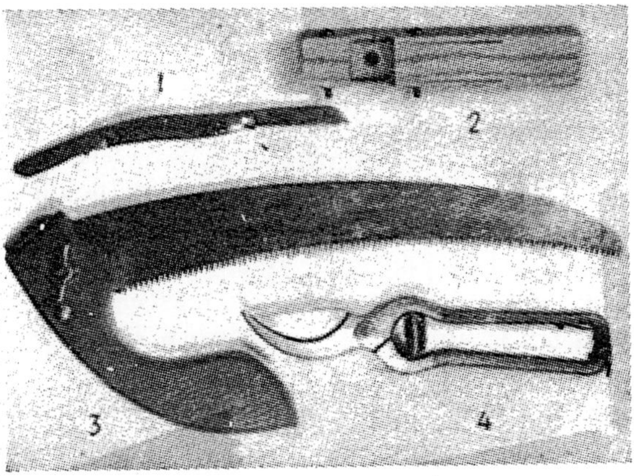

Fig. 112.—Tools for use in patch budding: (1) combination budding and grafting knife, (2) two-blade patch-budding knife, (3) folding pruning saw with crescent-shape blade, (4) pruning shears.

cut out to a depth of ½ inch. The length of the area cut out should not be so great as that of the safety razor blades. The blades are inserted into the two longitudinal cuts between the bolts and clamped securely in place by tightening the bolts.

Cleft Irons and Chisels.—The cleft iron or some modification of it is indispensable in cleft grafting. A popular form of cleft iron consists of a straight piece of steel with handle on one end, blade on one side, and a V-shaped shank at the other end on the opposite side from the blade. With the blade the stock is split down the desired distance and by use of the shank the two parts of the stock are held apart until scions are inserted in place. In the absence of cleft irons, wood chisels 2 inches wide may be used instead. Chisels are especially useful in making side bark grafts. Wooden mallets of convenient size are a necessity in connection with the use of cleft irons or chisels.

Pruning Saws.—In the different operations involved in propagation, several types of saws are essential. All of them have long, widely set teeth, for convenience in cutting green wood.

1. *Crescent Saws.*—A small saw with a curved blade 10 to 14 inches in length is very useful in pruning trees of medium size in preparation for top-working. Some of these saws have rigid handles, while in others the blade may be folded into the handle when not in use. It is held in open or closed position with a thumb nut. The teeth on these saws are set toward the handle, so that they cut on the return stroke.

2. *Tapered Saws.*—This is the designation for saws with a tapered frame, similar to those used in meat markets. The blades are removable, and a new one may be inserted whenever necessary. There is also the advantage that the blade may be set at an angle to the frame to permit cutting at difficult angles.

3. *Straight Saws.*—For removal of larger limbs, saws with straight blades of rather large size are most useful. A blade of 20 to 26 inches is sufficient for most work, but for some larger cuts a larger crosscut saw may be employed.

Pruning Shears.—The essential requirements of pruning shears are that they be strong and durable and make a smooth, clean cut with the least possible injury to the tree. Like all other tools, pruning shears vary in these respects. The amount and quality of work that an operator does is often an indication of the kind of shears he uses. Good shears need not be the most expensive, but they should be made of good steel.

1. *Hand Shears.*—These range in length from 4½ to 9 inches, and there are uses for which the different sizes are best adapted. The most common types of shears have a single cutting blade, operating against the side surface of a hook, or against a soft metal anvil. One of the best types, however, is the double-cut shear, in which there are two cutting blades. The springs may be of the volute type, which is probably the most common; double or single leaf type; or roller type. There are many individual preferences in shears, but the important considerations are ease and efficiency of operation.

2. *Lopping shears* are made with cutting blades similar to those in hand shears, except that the opening of the blades is

wider and the hook may have a slightly different shape. Handles of wood or steel, 20 to 24 inches long, give a greater leverage for cutting. These shears are very convenient in pruning work and are used also on berries, roses, and grapes. A newly introduced type of lopper has a patented slide shift that multiplies the pressure for cutting.

3. *Pole pruners* or *tree trimmers* have a hook-like cutting head mounted on a pole 4 to 18 feet in length. The cutting blade is

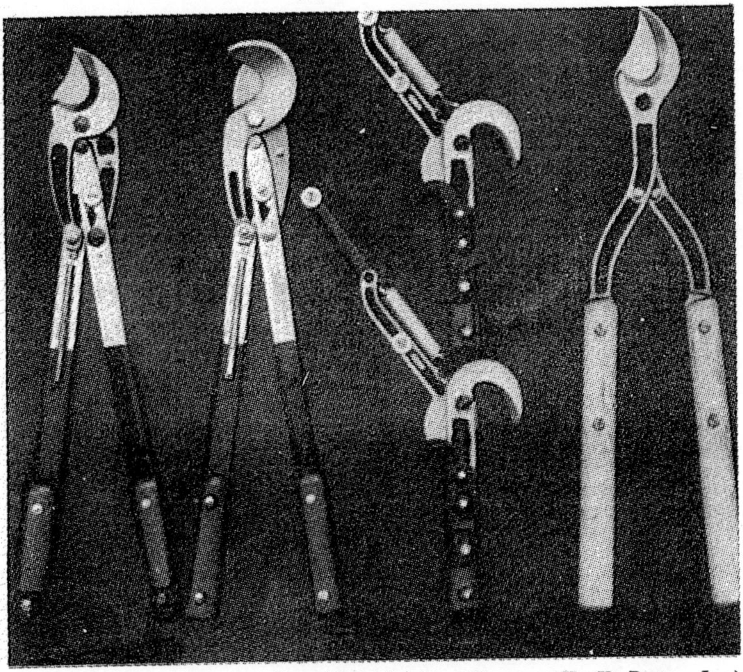

FIG. 113.—Different types of pruning shears. (*Courtesy H. K. Porter, Inc.*)

mounted so that it cuts upward against the hook. This hook is placed over a branch, and the shear operated by a downward pull on the rope, chain, or rod that is attached to the end of the cutting arm. This tool is useful in removal of small branches that are hard to reach with ordinary tools. It is especially valuable in cutting budwood from older trees, where much of the best wood is in the top.

Ladders.—In propagation work, the use of ladders is restricted to cutting of budwood, training of larger trees, and top-working.

In the last operation especially, ladders play a very important part. The use of ladders should always be encouraged in any work with trees. Branches are easily bruised and scraped by the shoes of workmen who climb in them.

1. *Stepladders* are useful in trees of moderate height. The 6-foot and 8-foot ladders are not difficult to move about and afford a steady footing. For nursery or orchard work, stepladders should be made with a single supporting leg, attached at the top. Such ladders may be placed close in to the tree without injury to the branches, and they are more easily adjusted on uneven ground.

FIG. 114.—Stepladder of a desirable type for nursery and orchard work.

2. *Straight ladders* may have either the ordinary open top or a closed top which tapers to a point. Such ladders should have a flaring base to give greater stability. These ladders are made of light materials in lengths up to 22 feet, which is about as large a ladder as one man can handle. Extension ladders reaching up to 30 feet are used in some cases in top-working large trees. The rungs of very long ladders are spaced wide apart so as to make them lighter and easier to handle.

3. *Hook ladders* are designed especially for use in tall trees. They are useful in operations involved in top-working very large pecan trees, for example, and also in the harvesting of the crop. These ladders are relatively small and light and are equipped with a strong iron hook. Workmen carry the ladder into the tree with them and by placing the hook over various large limbs are able to climb progressively to the topmost parts of a tree with relative ease.

Wound Coverings and Protective Materials.—Pruning wounds of large size, and other wounds resulting from injury, should be covered with some protective material, primarily to prevent infection by disease. Many materials that have been used in the past have injured the bark and in this way delayed healing. Wounds resulting from the removal of small branches, especially those 1 inch in diameter or less, will often heal before any rotting of the wood occurs. In damp climates, however, and in regions

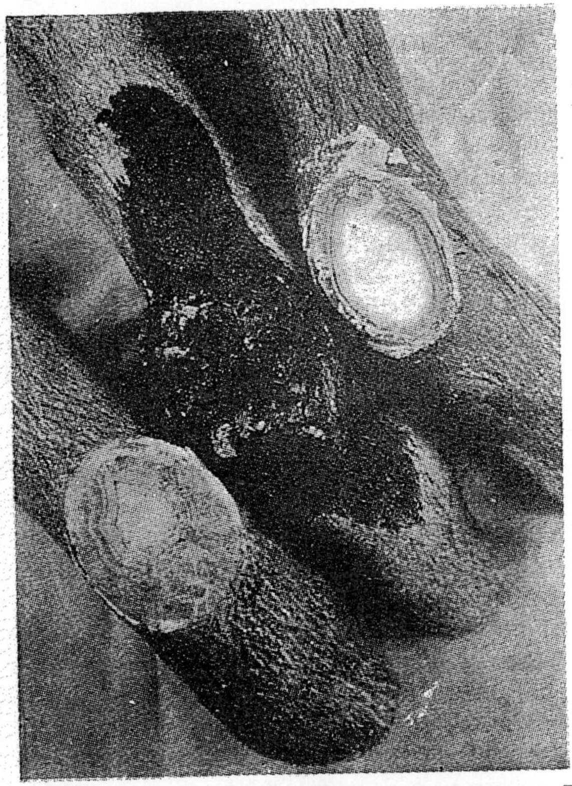

FIG. 115.—Wound healing as influenced by different treatments. Top, petroleum varnish, zinc oxide, yellow oxide of mercury compound. Center, abnormal growth caused by roofing cement containing coal tar. Bottom, untreated wounds may overwall but the exposed wood checks and allows moisture to accumulate in the wound. (*Courtesy Texas Agr. Exp. Sta.*)

where disease is prevalent it is advisable to treat all cuts as soon as the surface has become dry.

1. *Wound Coverings.*—The requirements of a good material for this purpose are (1) it should give protection against infection, adhering to the surface and neither cracking nor running, and

(2) it must not be injurious to plant tissue or result in shrinking of the tissue or excessive callus development. Often it is advisable to make two different applications to accomplish the dual object of disinfection and continued protection. Good materials for disinfection of wounds are plant-safe carbolineum or bichloride of mercury (alcoholic solution) combined with pine rosin. The second coat may be made of asphaltum carbolineum, which is black; or a white enamel of zinc oxide and petroleum varnish with mercuric chloride may be used. The white compounds are affected less by exposure to the sun than are the black ones. The materials described have been used with good results in treatment of pruning wounds on citrus trees in Texas; *Citrus* requires protection against disease infection and thus serves as a good test for the materials.

2. *Rabbit Repellents.*—Damage to fruit trees in the nursery or in the early life of the orchard by rabbits is reported from many different sections. Nearly all the common fruit trees are attacked in winter, when the food supply is limited by snow. The animals eat the bark of the trees up as high as they can reach, so that recovery or repair of the tree is impossible.

Various *mechanical* means of protection are used effectively in combating rabbits and other rodents. Heavy paper or old newspaper wrapped around the trunk to a height of 18 to 20 inches is often used, but has the disadvantage of providing overwintering quarters for certain insects. Cylinders of veneered wood or hardware cloth provide the same kind of protection, but are more expensive. Paper wrappers or veneered cylinders have the advantage of discouraging attacks by flat-headed appletree borers because these do not readily attack shaded trunks. *Chemical* repellents have been used for some time, but some of these have done more damage than the rodents might have done. One compound that has been used successfully in Michigan and other states is made as follows:

Rosin	35 pounds
Fish oil	15 pounds
Copper soap (copper oleate)	15 pounds

The ingredients are melted over a moderate fire and mixed by stirring. The compound is applied while warm by means of a brush.

Questions

1. Why are grafting waxes and similar materials used in budding and grafting?

2. What are the essential characteristics of a good grafting wax?

3. How is alcoholic wax made? Hard wax? Hand wax?

4. What precautions should govern the use of paraffin in graftage?

5. What various materials are used for tying buds? Grafts?

6. What are the merits of wax string when used for tying whip grafts that are made indoors? What are the merits of waxed or medicated tape when used for the same purpose?

7. What characteristics make rubber bands popular for tying buds?

8. Name the different types of pruning shears. Give principal uses of each.

9. Describe the different kinds of ladders used in propagation work. Suggest principal uses of each.

10. What compounds are used to protect wounds? What essential characteristics should such compounds have?

11. Tell how to protect young trees against rabbits and other rodents.

Suggested References

1. FRIEND, W. H.: Wound Protectants for Orchard Trees, *Tex. Agr. Exp. Sta. Progress Rept.* 587, 1939.
2. MACDANIELS, L. H.: Tree Bracing and Wound Treatment in the Orchard, *N. Y. (Cornell) Agr. Ext. Bull.* 312, 1935.
3. NIELSON, J. A.: Condensed Information on Rabbit Repellents, *Mich. Agr. Exp. Sta. Quart. Bull.* XVI, No. 2: 59–62, 1933.
4. TALBERT, T. J.: Top and Double Working, and Bridge Grafting of Fruit Trees, *Mo. Agr. Exp. Sta. Circ.* 196, 1938.
5. WORTHLEY, H. N., and L. C. MARSTON, JR.: Preparing Chemically Treated Bands for Codling Moth Control, *Pa. Agr. Exp. Sta. Bull.* 330, 1936.
6. YOUNG, H. C., and PAUL E. TILFORD: Tree Wound Dressings, *Ohio Agr. Exp. Sta. Bimonthly Bull.*, XXII, No. 186, 1937.

CHAPTER XII

METHODS OF GRAFTING

Grafting is the operation of inserting a portion of a limb of one plant into another in such a way that the two unite. The methods of grafting differ with respect to (1) the season at which the operation is done, to (2) the technique involved in fitting the scion and stock together, and to (3) the adaptation to different kinds and sizes of plants.

Cleft Graft.—The cleft graft is one of the oldest methods of grafting. It is also one of the simplest forms and may be used with good success by amateurs on trees that are easily propagated. The principal use of the cleft graft is found in top-working large trees where limbs from 1 to 3 or 4 inches in diameter are to be grafted. Cleft grafting should be done during the latter part of the dormant season just before active growth starts in the springtime.

With a sharp saw the limbs are cut off squarely at a place where they are straight and free from crooks and side branches. As far as possible, limbs should be cut off at such places that when grafts are inserted they will develop into a symmetrical top of good conformation. The end of the stub is then split with the cleft iron and is ready for the insertion of the scion. Scions are prepared by cutting the end in a wedge shape, with one edge of the wedge slightly thinner than the other. It is customary to make the cuts so as to leave a vigorous bud just above the top of the wedge and on the same side as the thick edge of the wedge. The cuts forming the sides of the wedge should be long, to insure gradual rather than abrupt tapering toward the apex of the wedge. They are best made with a single full stroke of a sharp knife, and each cut should be a perfect plane. When prepared in this way, scions will fit into the cleft of stock better than if cut in a haphazard way or if the wedge tapers too abruptly. Two or three good buds on the scion above the crown of the stock are sufficient; the cut on the upper end of the scion should be made ⅛ inch above a bud. One scion is enough for each stock less than 1 inch in diameter, but it is a common practice to insert two scions in larger stocks—one in each side of the cleft. On sloping branches the cleft should be made horizontally, so that the scions are lateral to each other and not one above the other. In adjusting a scion, the cambium

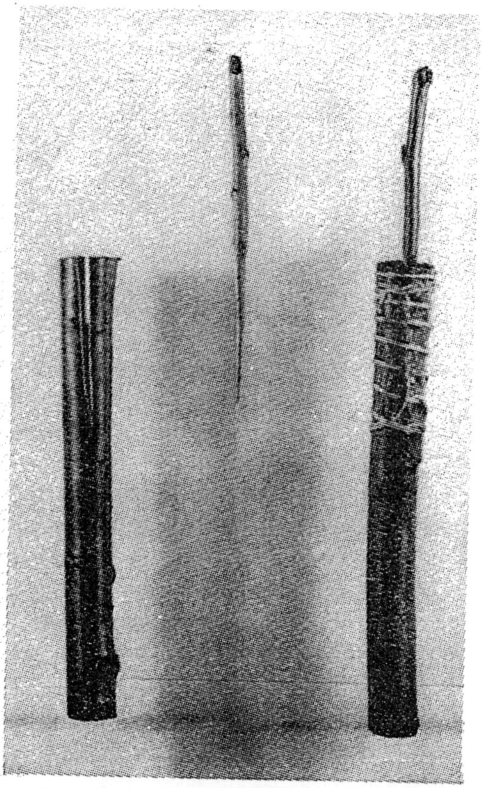

FIG. 116.—Cleft graft showing stock, scion, and completed graft.

layer should be placed in contact with that of the stock at as many points as possible. The thick edge of the wedge of the scion should be toward the outside, so that the pressure exerted by the two halves of the stock will be at points where the cambium layers coincide. If the scions are not held securely in place the pressure of the two halves of the limb should be supplemented with strong cord. The end of the stub and the entire length of the split are covered with grafting wax.

In the event two scions grow in one stock the weaker one is kept headed back for a year or two and is finally cut off entirely even with the crown of the stock. In the meantime it will assist in the formation of new tissue over the end of the stub. Two scions are likely to form a weak, narrow V crotch if both are allowed to grow permanently in one stock.

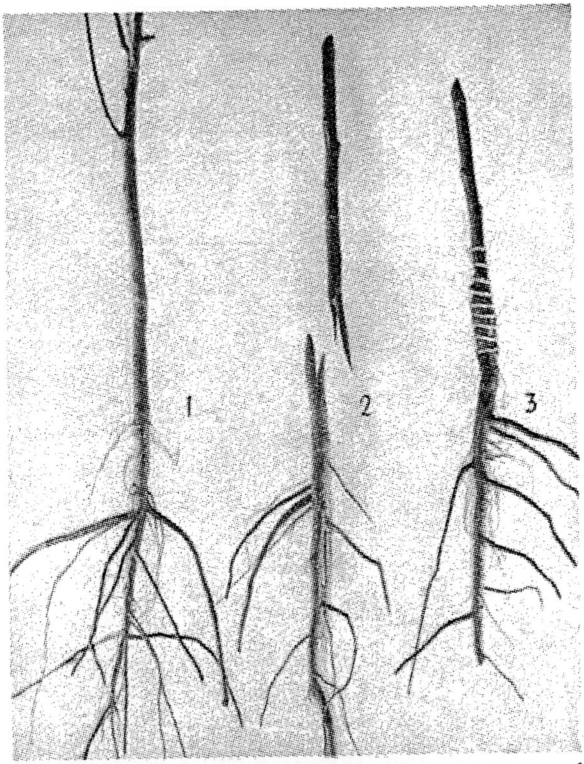

FIG. 117.—Detail of whip graft. (1) Stock, (2) stock and scion cut, and (3) graft completed and tied.

Whip Graft.—Whip grafting has long been a popular method of propagation and is the means whereby many different kinds of plants are reproduced. The method is used on stocks that are relatively small; those larger than ¾ inch in diameter cannot be conveniently whip-grafted because of the difficulty involved in making the cuts properly.

The top of the stock is cut off with a diagonal cut that should be about 1¼ inches long on a ½-inch stock. This cut should be proportionately longer on larger stock and shorter on smaller stock. The scion, having been cut previously to a length of 5 or 6 inches, is cut across the lower end in a similar way to the stock. The cut on the stock should be made upward and the cut on the scion should be made downward. Each of these surfaces should be smooth and as nearly a plane as possible; uneven or wavy surfaces prevent proper contact between the two cambium layers. A second cut is required on both stock and scion. It should begin one-third of the distance from the apex of the cut to the base, and extend toward the base. This cut, which forms a tongue, should be slightly across the grain of the stock, its course being toward the base of the original slanting cut. When prepared in this way, the stock and scion are snugly fitted together, the two tongues interlocking. Extreme care should be exercised to have the cambiums of the two in contact with each other on one side. If the scion and stock happen to be of the same size, it is possible to have the cambiums coincide on both sides, in which case the chances for a successful union are increased. Little trouble, however, is experienced in obtaining a union if the cambium layers meet on one side only.

It is not customary to apply wax to whip grafts. They are commonly wrapped securely with waxed string. Cotton twine may be used, but it requires tying, whereas the end of waxed string is held in place by the wax. Waxed tape and medicated nursery tape are becoming increasingly popular as wrapping materials for whip grafts because they discourage the formation of callus knots.

Whip grafting may be done in the nursery, or the stocks may be dug and grafted indoors. This latter practice is known as *bench grafting*. Such grafts are usually packed in moist insulating material, stored in a room where the temperature is from 75 to 85°F. for 1 week or 10 days for callusing, and then planted in the nursery. If planting is delayed, the grafts can be held in cold storage until needed. In either nursery grafting or bench grafting it is customary to place the graft on the portion of the stock just below the ground line, and in replanting bench grafts the union should be planted slightly below the ground level.

Bench grafting may be done successfully any time during the dormant season if a favorable temperature for callusing is provided. Nursery grafting should be delayed until the near approach of the growing season, because the temperature in the nursery is not likely to be favorable for callusing during the winter.

Bark Graft.—The bark graft is unique in that the period during which it is done is after the bark begins to slip in the spring rather

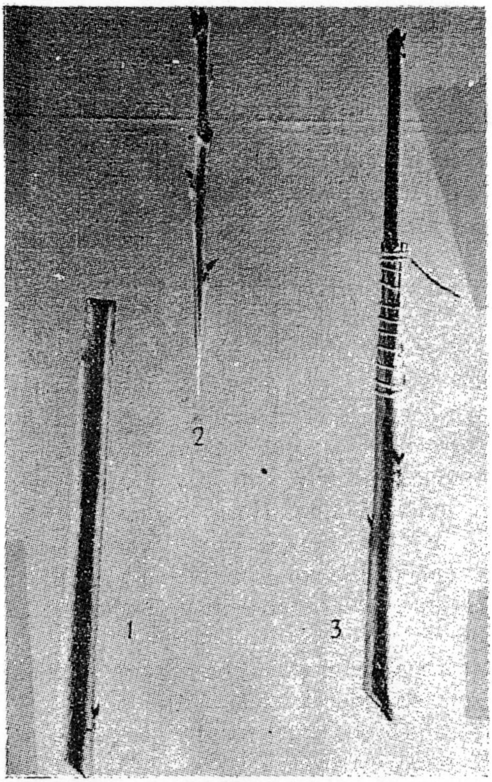

FIG. 118.—Steps in inserting a bark graft. (1) Stock, (2) scion, (3) scion inserted and tied.

than during the dormant season. Scions used in bark grafting, however, should be thoroughly dormant. Stocks for bark grafting range from ½ to 4 inches in diameter. Scions may be from ⅓ to ¾ inch in diameter, depending on the size of stock on which they are to be used. With reference to position on the stock, bark grafts may be *terminal* or *lateral*, the former being used more commonly.

For a terminal bark graft, the stock is sawed off squarely, at a point where the bark is smooth and free from knots. The bark is split downward from the crown about 2 inches. If the bark is thick and rough, it is pared down so as to render it flexible on each side of the downward cut. On the lower end of the scion a sloping, downward cut is made, the surface of which should be 2 or 3 inches long, depending on the size of the scion. On the side of the scion opposite the lower portion of the cut just described, two additional cuts are made. They are each about 1 inch in length and are designed to expose additional cambium cells, which make contact with similar cells of the stock and increase the chances of a union. These cuts intersect the first cut and result in a distinctly pointed scion, which may be readily inserted. Scions are usually 4 or 5 inches long and should contain two or three good buds. As soon as the scion is prepared it is inserted beneath the split bark of the stock. The long cut is placed in contact with the xylem or wood. The scion should be pressed down so as to allow only a small part of the cut to be exposed above the crown. Two or three scions may be placed in large stocks, and the chance of growth is increased. In the event more than one grows, only the most desirable one is allowed to remain permanently.

Tying with durable string and the application of wax complete the bark graft. Ordinary cotton twine may be used in tying on small stock, and stronger cord for larger stock. The tying material may be left on as long as it does not girdle, in order to support growing scions. Some propagators make a practice of placing a twig 3 or 4 inches long and the size of a pencil on each side of the inserted graft, and outside the flaps of bark which extend over the scion. In tying, the string is pulled over these two twigs. The effect is to press the bark of the stock in closer contact with the scion, and decrease the air space on each side of the scion. The practice is especially recommended for large stocks, the bark of which has a tendency to bulge away from the stock when grafts are inserted beneath the bark. Small nails may be used to hold bark grafts securely in place instead of twine. Cigar-box nails are about the right size and two are sufficient for each scion. Any good grafting wax may be used satisfactorily in excluding air and water from bark grafts. Melted paraffin is also used.

The essential difference between a terminal and a lateral bark graft is in position. There is little difference in the technique involved. A cut is made on the stock at a right angle to the main axis of the stem to a depth of ½ to ¾ inch at a point where the bark is smooth and will permit the easy insertion of the graft. With a wood chisel or large knife the stock is notched above so as to intersect this cut. This notch permits easier insertion of the scion and better contact of it with the stock. The bark is split downward from the first cut made and a scion inserted. Details of preparing the scion and its insertion,

FIG. 119.—Cross section showing union of bark graft after one season of growth. Note that it formed union on one side only.

tying, and waxing are the same for lateral grafts as for terminal. Foliage is left on the stock above lateral grafts, and this is thought to enable them to make better unions than terminal grafts. The rate of growth of lateral bark grafts can be regulated by the extent to which the native foliage beyond the graft is cut off; advantage is taken of this in preventing shoots from growing too vigorously and becoming top-heavy.

Inlay Graft.—The inlay graft is similar to the bark graft in many respects. The size of stock on which it is used, season, and graftwood are the same, but the graft itself is made differently. One large sloping cut is made opposite a good bud on the scion. This cut is to fit next to the xylem of the stock. On the side opposite the lower part of this first cut a second one that exposes additional cambium is made. The long cut surface of the scion is then placed on the stock, and a strip of bark the width of the lower part of scion is cut on each side. This flap is then peeled down and the scion inserted so that its long cut surface will be next to the xylem of the stock. The top part of the flap is cut off and the lower part is pressed over the outside

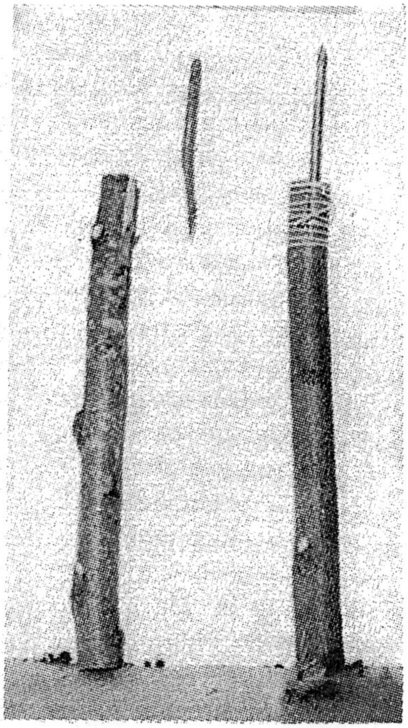

FIG. 120.—Inlay bark graft.

cut surface of the scion. It is then tied and waxed. It is thought that inlay grafts make a stronger union than bark grafts. A double inlay graft is commonly used for bridge grafting, which is described later.

Other Methods.—The *veneer graft* is used chiefly in propagating (1) coniferous evergreen plants that are difficult to graft, and (2) herbaceous greenhouse plants. A long sloping cut is made in the top or side of the stock, and in it the scion, which has been cut wedge-shaped on the lower end, is inserted. When side cuts are made the top of the stock remains intact until union is established, after which it is removed.

Cutting grafts are made by grafting a scion onto an unrooted cutting. Such grafts are usually stored as soon as they are made, under favorable temperature and moisture conditions, until the scion and cutting stock unite by callusing. They are then ready for field planting. If necessary they may be held for a few days or weeks properly insulated in cold storage before planting. Cutting grafts are sometimes planted in the field as soon as the scion and unrooted stock are joined together, in which event the two are expected to unite by callusing during the same period that the cutting-stock is forming roots.

Root grafting refers to a grafting operation whereby a scion is grafted onto a root, used as a stock. In most cases whole root systems of small plants except for slight pruning are used; the scion is inserted in the upper part of the taproot, resulting in *whole-root* grafts. If one root is cut into two or more pieces and each is used as a stock, the grafts are called *piece-root* grafts.

A *nurse-root graft* is made by grafting a long scion onto a relatively short root piece. It is usually done indoors, and when the callused graft is planted in the field the graft union is set at a depth that permits the top of the scion to extend barely above the ground level. The nurse root supports it for a time, but because of its depth seldom makes continued or extensive growth. In the meantime roots are expected to form on the lower portion of the scion, in which case the plant becomes own-rooted, and the nurse root may or may not be cut off at a later date. When dwarfing species are used as rootstocks for nurse-root grafts, it is seldom necessary to remove the rootstock after the scion develops roots. Other means of inducing rooting of the scion involve the use of an inverted nurse root and girdling with string or wire at the point of union. A nurse root that is susceptible to the attack of a disease or insect pest may sustain a scion temporarily; if the scion is resistant to the trouble, roots that are formed on it will gradually replace those of the susceptible rootstock. Plants that root poorly from the scion may root satisfactorily from *scion shoots* that are caused to grow from below the soil surface by deep planting of the graft. Some plants are valuable for rootstocks because of having a good root system. If seeds do not reproduce a uniform rootstock of such plants, and if they cannot be grown successfully from cuttings or layers, the nurse-root graft is a possible method whereby they can be produced on their own roots.

Aftercare of Grafts.—Grafts in small stock usually require some training to cause them to grow into trees of the desired shape or form. Frequently two or more buds on a graft grow, with the result that a tree of undesirable shape is produced. The weaker growing bud should be cut off and all growth forced into one bud at first. After it has formed a single standard of desired height, lateral branches may be allowed to develop. Some plants are more valuable if they are branched, and these are trained accordingly. In order to insure straight, upright growth of grafts, some propagators follow a practice of staking the trees with a lath.

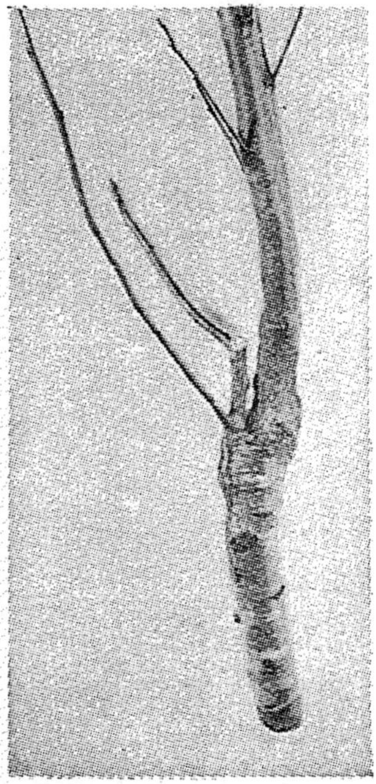

Fig. 121.—Cleft graft with one scion allowed to grow and the other retarded in growth. The growth of the smaller scion at left aids in healing the stub.

Grafts in large stocks usually make a very vigorous growth during the first one or two seasons. They are likely to become top-heavy and to be blown down or broken off by high wind. This often happens, even where the stock and scion have made a good union, but is more likely to occur if the union is poor or defective. Large stocks are rather inflexible, and the pressure exerted by wind against a graft is greatest at the point where the stock and scion meet.

Danger from loss of grafts by breaking can be lessened by bracing and by pruning. A lath or board tied securely or nailed to the stock in a way that will permit it to extend beyond and serve as an anchor to which the growing graft may be tied is a desirable method of protecting grafts from high winds. Grafts that make unusual length growth may be headed back in order to reduce the strain at the point of union. It is especially important that this be done on fast-growing grafts before they begin the second season of growth.

Stock in which grafts are inserted should be covered with a protective coat of a substance that will keep moisture from soaking into the exposed cut surface. Ordinary white lead paint is

FIG. 122.—Lateral wound on apple showing bridge grafts which have united successfully and completed one season of growth. (*Courtesy Dept. of Hort., Mich. State Coll.*)

sometimes used, but the oil in it soaks into the wood, leaving the powdered material, which affords little protection, on the wound surface. Orange shellac is also used and is quite satisfactory. Various commercial preparations are available that, when applied properly, afford protection from moisture and tend to minimize the injurious effects of decay-producing organisms.

Squirrels frequently do considerable damage to young grafts of pecan trees by gnawing the tender bark and partially or completely girdling the limbs. Guards made of tin, placed around the trunk of the tree, prevent squirrels from reaching the grafts, provided no other trees are near by.

Bridge Grafting.—Bridge grafting is not a means of propagation in the sense that the other methods are. It is a means of repairing tree wounds caused by cultivating implements and by mice, rats, rabbits, and other rodents. Most injury caused by animals is done during the wintertime, when other plants on which they feed are scarce. These pests are especially troublesome when snow is on the ground and their food supply is even more limited. It is not uncommon for trees to be partially or completely girdled from the ground to a height of 12 or 15 inches. Trees from nursery size to those 3 or 4 inches in diameter are subject to their attack. They do not seem to have marked preference for any one kind of trees; in one instance, rabbits injured trees of peach, plum, apple, pear, and pecan, seemingly without favor. Field rats have been known to do serious damage to bearing fig trees by girdling the trunks. Perhaps the wisest course to follow in combating such damage is to adopt preventive measures. Precautions should be taken to avoid injuries that result from cultivation, such as plow cuts and bruises. Clean cultivation discourages habitations of rodents in the orchard and thereby lessens the chances of damage. When injury is anticipated, protection may be provided by the mechanical means or chemical treatments discussed in the previous chapter.

Trees that have been injured to the extent of having all the tissue outside the xylem destroyed, even for a short distance, are said to be *girdled.* Those that are completely girdled die of root starvation sooner or later unless the connection between the root system and supporting top is reestablished. Some kinds of trees callus-over wounds, either by regeneration or overwalling, and rebuild the destroyed phloem tissues before their root system ceases to function. Elm trees are difficult to kill by girdling because of their ability to callus-over wounds rapidly. A pecan tree is known to have callused-over about 5 feet of girdled area, but such instances are extremely rare.

Fruit trees will seldom callus-over large wounds that completely encircle the trunk.

Sap flow between the upper and lower margin of a girdled area may be reestablished by causing congenial scions to unite with the tree both above and below the girdle, in such a way as to bridge the gap. This operation is called *bridge grafting.* It should be done in early spring about the time the injured tree starts growth. Only thoroughly dormant scions are used. The irregular edges of the girdled area should be cut back evenly

FIG. 123.—Mango seedlings being grown in wooden pots preparatory to inarching.

to fresh tissue preparatory to inserting the bridge grafts. The scions are cut at each end; after the matrix is prepared above and below the injury on the stock, as for the inlay or the bark graft, they are inserted, tied, and waxed. Large wounds may require more than one bridge graft. It is customary to space them 2 or 3 inches apart. After they are united they transport elaborated sap past the injured area to the roots.

Approach Grafting.—The approach graft or inarching is a special form of grafting in which the scion unites with the stock while it is still attached to the parent plant. It is therefore necessary that the plants to be used for stocks and as a source of scions, respectively, grow close together. The stock may be moved in a clay pot or it may be transplanted bare-rooted to the vicinity of the tree from which scions are to be obtained. On both the stock and scion a long cut is made through the cambium and slightly into the wood. The cut surfaces are brought

FIG. 124.—A living brace designed to strengthen the weak crotch shown at *X*. The lateral limb arising at *Y* was caused to unite with the limb shown on the right at *Z*. A spiral inarch was used; it was made in April, and the picture shows it as it appeared the following winter.

together, and the stock and scion are tied firmly and waxed. After the two have united, the scion in some cases is severed below the union and the stock above that point, resulting in a new plant that is composed of a rootstock and a grafted top. In other cases, two parts of a plant, or parts of two plants, are caused to unite and grow together permanently and neither component is severed below the point of union.

The technique of either the bark graft or the inlay graft may also be used for approach grafting. Though the approach graft is tedious and cumbersome, it has a variety of important uses:

1. *Difficult Species.*—Approach grafting, or inarching, is used in the multiplication of plants that are extremely difficult to propagate by other asexual means. It may be used, for example, in propagating the Muscadine grape, which responds very

FIG. 125.—Living braces of apple created by twisting branches from adjacent limbs together and allowing them to grow together. (*Courtesy Dept. of Hort., Mich. State Coll.*)

poorly to ordinary methods of graftage, and is also used in propagation of avocado and mango.

2. *Novelties.*—Approach grafting is used to some extent, principally in European countries, as a means of providing novelties for ornamental gardens. At one place in Italy the tops of oaks on each side of an avenue have been grafted together by means of inarching.

3. *Bracing.*—Living braces may be provided in the framework of a tree by grafting a small side limb from one branch into an adjacent branch. The same protection may be provided by twisting together small limbs from two branches and letting

them unite naturally. These are valuable in strengthening limbs that form V or otherwise weak crotches. Living braces grow and become stronger as the tree becomes larger, and may ultimately provide a more effective support than could be provided in any other way.

FIG. 126.—The two small trees on either side of the center one were inarched into it above the girdled area, shown at *X*.

4. *Repairing Trees.*—Approach grafting may also be used to repair trees that have been girdled or injured in some other way near the ground line. Young trees are planted about the base and their tops are inarched into the trunk above the injured area. These ordinarily make a very rapid growth and ultimately become a permanent part of the old tree. A sprout that arises from below a wound on the trunk of a tree may be inarched

into the trunk above the wound; when it unites, it provides a connection that will cause more even and rapid healing of the wound.

Questions

1. Define grafting. Distinguish between grafting and graftage.

2. When is the season for cleft grafting? What size stocks are suitable for the cleft graft?

3. Describe the steps in preparing the stock and the scion for the cleft graft.

4. What are the objects of inserting two scions in one stock? What procedure is followed in the event both grow?

5. Describe the steps in preparing the stock and the scion for the whip graft. What size stocks are suitable for this method?

6. What is the distinction between bench grafting and nursery grafting? When is the season for each?

7. Tell how to prepare the stock and scion for the bark graft. Distinguish between a terminal and a side graft.

8. Describe the inlay graft.

9. When is the season for bark and inlay grafting? Why is it different from the season for other methods of grafting?

10. What treatments are involved in the aftercare of grafts?

11. What are the chief uses of the veneer graft? Describe the cutting graft.

12. What are the different uses of the approach graft?

13. What various treatments are prescribed for trees that have been girdled?

Suggested References

1. Bryant, L. R.: Plant Propagation, *Col. Agr. Exp. Sta. Bull.* 441, 1938.
2. Beaumont, J. H., and Ralph H. Moltzau: Nursery Propagation and Topworking of the Macadania, *Hawaii Agr. Exp. Sta. Circ.* 13, 1937.
3. Hodgson, R. W.. Saving the Gophered Citrus Tree, *Calif. Agr. Exp. Sta. Circ.* 272, 1923.
4. Jacob, H. E.: Propagation of Grapevines, *Calif. Agr. Ext. Serv. Circ.* 101, 1936.
5. Lewis, I. P.: Grafting and Budding Fruit Trees, *Ohio Agr. Exp. Sta. Bull.* 510, 1932.
6. MacDaniels, L. H.: Tree Bracing and Wound Treatment in the Orchard, *N. Y. (Cornell) Agr. Exp. Sta. Ext. Bull.* 312, 1935.
7. Reed, C. A.: Nut-Tree Propagation, *U. S. Dept. Agr., Farmers' Bull.* 1501, 1926.

CHAPTER XIII

METHODS OF BUDDING

Budding is a form of graftage in which only the buds at one node of a scion are inserted in a stock. Buds are cut from scions in such a way as to have a relatively small amount of bark surrounding them; and in some methods a small sliver of the wood beneath the buds is included also. Buds are customarily placed on the stock at internodes; it is not necessary, in fact is less desirable, to insert them at places where native buds grew on the stock.

Shield, or T, Bud.—The most widely used method of budding is the shield, or T, bud. It is known as *shield budding* because the bud, when cut from the bud stick, resembles a shield in shape and as *T budding* because the two cuts made on the stock intersect so as to form a T.

A perpendicular cut 1 inch long, or less, is made on a smooth portion of the stock. It is followed by a horizontal cut across the top at right angles to it. These two cuts extend only through the bark and should not go into the wood.

The bud is cut by starting ½ inch below it and cutting upward and obliquely inward to a point about ½ inch above the bud. The knife is then withdrawn and a horizontal cut through the bark only is made ½ inch above the bud. With a sidewise twist the bud is next gently removed from the bud stick without wood adhering; this practice is known as *flipping* or *popping* the bud. The pointed lower portion of the bark is inserted underneath the two flaps of the T cut on the stock, which may have been loosened with the knife. The bud is then pushed downward until the top of it is well below the horizontal cut on the stock. When the bark of the stock slips readily the bud can be pushed into place with little difficulty.

Buds may also be cut in about the same way as just described, except that the horizontal cut extends into the wood, which is transferred along with the bud. If the bark on the scion will not slip, it is necessary to cut the bud with some wood adhering.

If the bark of the stock does not slip well, it may be desirable to cut the bud so as to include the sliver of wood, since it makes the bud more rigid and easier to insert in the stock. Under no circumstance should an individual bud be cut from the scion until the stock has been prepared for its reception so that the transfer can be made immediately.

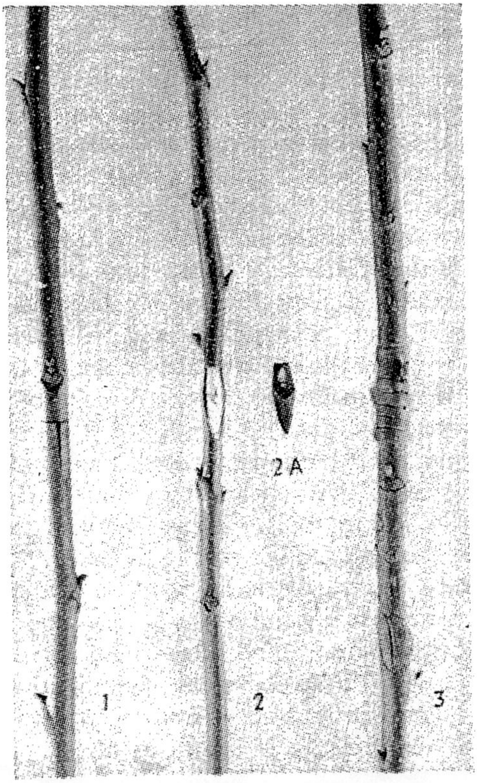

FIG. 127.—Showing successive steps in T budding. (1) Stock prepared for reception of bud; (2) bud stick, indicating technique of removing the (2A) bud; (3) bud inserted in stock and wrapped with rubber budding strip.

During the seasons when, or in regions where, rainfall is excessive, *inverted T buds* are frequently used. The horizontal cut on the stock is made at the lower end of the perpendicular cut, and the bud is cut from above rather than below. The bud may be cut with or without wood adhering, but it can be placed more easily if wood is included. It is inserted by being pushed upward beneath the flaps of the inverted T cut, which serve to shed water.

Moist raffia, strips of corn shucks that have been moistened, waxed tape, cotton twine, and rubber budding strips are used for tying T buds. If the tying is done carefully, it is not necessary to apply wax. Stocks to be T-budded should not be greater than 1 nor less than ¼ inch in diameter. Those the size of a lead pencil or slightly larger are considered ideal.

FIG. 128.—Citrus stock cut back to bud union and budded top staked.

The T, or shield, bud is the most popular of the methods of budding. It is used commonly for propagating plants such as roses, peaches, apples, and citrus that are easy to bud; but it is seldom used on species that are difficult, such as the pecan or walnut.

Shield, or T, budding is done successfully at any season of the year when the bark of the stock will slip freely. By one practice, buds are inserted in late summer; they unite with stocks but remain dormant over winter and are forced into growth the following spring. This practice is referred to as *dormant budding*. After the buds grow one season, the trees that they produce are known as *one-year-old* trees, though the rootstocks are two years old. Buds may also be set in early summer and forced into growth shortly afterwards. This is commonly known as *June budding*, because much of the budding is done during the month

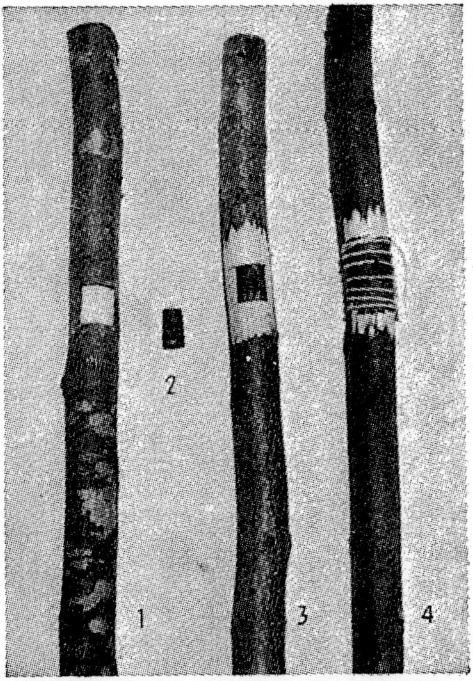

FIG. 129.—Showing steps in inserting a patch bud in thick bark. (1) Stock with square removed; (2) bud; (3) bud in place; (4) bud protected with waxed patch and tied with cotton twine.

of June. In regions where spring comes early and the growing seasons are long, June-budded plants ordinarily become large enough by the end of the first growing season to be planted out in the orchard.

Patch Bud.—The patch bud is a popular form of budding in the propagation of species that are rather difficult to propagate. Essentially, the patch bud method consists of removing a square or rectangular piece of bark from the stock and inserting in its place a bud of a desired variety on a similarly shaped piece of bark. Two parallel cuts, ¾ or ⅞ inch apart, are made on the stock, preferably with a two-bladed budding knife. The cuts should be made perpendicular to the stock, and should be about 1 inch long. With a sharp pocket knife, two longitudinal cuts are next made. They likewise should be about ¾ or ⅞ inch apart and each should intersect the two horizontal cuts, resulting in the square or rectangular "patch." Similar cuts are made above, below, and on each side of a bud on a bud stick with the

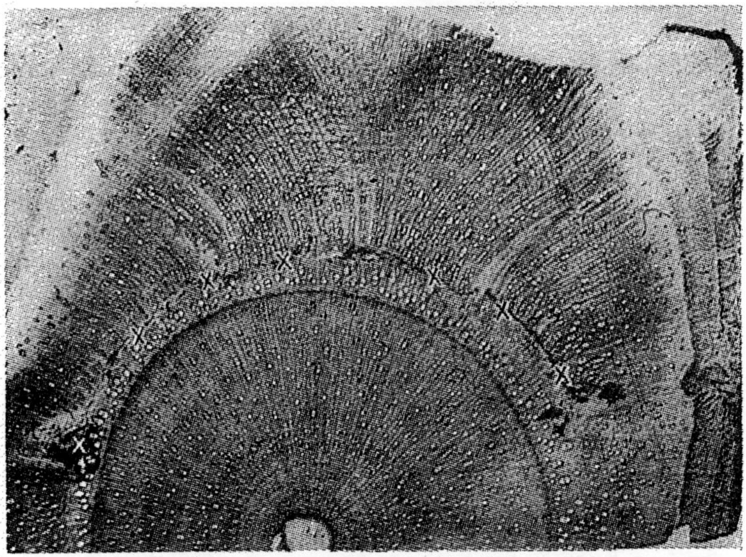

FIG. 130.—Cross section of pecan stem through patch bud that had united and grown part of one season. Union of the patch and stock took place along the line indicated by *X*.

same tools that were used on the stock. Care should be exercised in removing the bud from the budwood in order to avoid splitting the bark beneath the bud. The bark should be lifted carefully on one side, or both sides if necessary, and the bud loosened by a lateral twist. The bud is held in place on the scion while the patch of the stock is flipped off, and the bud is then quickly transferred to its place.

In making the transfer it is important that the delicate cambium cells on the underside of the bud and on the exposed surface of the stock be subjected as little as possible to mechanical injury and exposure to air. The bud should fit snugly in its new location and should be tied immediately. Cotton twine and waxed tape are commonly used for tying these buds. All cut surfaces (joint lines) should be covered with good grafting wax, melted paraffin, waxed cloth, or some similar material.

Stocks that range in size from ½ to 4 inches in diameter may be patch-budded quite successfully. For the larger ones

FIG. 131.—Showing one season's growth of patch buds inserted in rough bark of pecan. The shoots are braced to prevent them from being broken by wind. (*Courtesy Tex. Agr. Exp. Sta.*)

it is usually necessary to pare the rough outer portion of the bark down to the thickness of the budwood bark at the time the bud is put in place. This precaution is essential to the success of the inserted bud in that it allows the pressure of the tying material to be exerted on the bark of the bud rather than on a thick shoulder of bark on either side of it. Buds for large stocks should be selected with special care. They should be taken from smooth, straight bud sticks and only large, plump buds should be used; small buds are difficult to force and should be discarded.

Patch buds may be inserted successfully at any season of the year when the bark will slip freely. Those that are set early in a season are usually forced, while the ones that are set late remain dormant over winter and are forced the following spring.

There are several other methods of budding in which the buds unite with the stock in much the same manner as the patch bud and differ from the patch bud only in minor details. These are conveniently considered in connection with patch budding.

1. *Ring Bud.*—Ring budding differs from patch budding in that a cylinder of bark is removed from the stock in order to form a matrix; and the bud, when placed, extends nearly if not all the way around the stock. The stock is completely girdled, and, if the bud fails to unite, the top part of the stock ultimately dies. The nature of this method renders it impractical except for small stocks, those no more than ⅔ or ¾ inch in diameter.

2. *H Bud.*—A modified patch bud known as the *H bud* differs from the patch bud in that the two horizontal cuts on the stock are intersected by only one longitudinal cut. The two flaps of bark on either side of the longitudinal cut are lifted slightly and the bud patch is inserted underneath, from above or below. In preparing the bud, the two horizontal cuts are made and the sides cut so as to form a square; but the longitudinal cuts are cut at an angle of about 45 degrees, to enable the bark of the stock to fit over them more snugly.

3. *Skin Bud.*—The skin bud is a recent addition to the arts of plant craft. The matrix on the stock is prepared by cutting just through the bark, but not into the wood, in such a way as to remove an oblong piece of bark ⅔ inch long. The size of stock determines, in a measure, the width of the oblong piece of bark removed. In a similar way the buds at a node are cut, beginning below the bud ⅓ inch, cutting inward, then upward between bark and wood, and out again ⅓ inch above the bud. The thin sliver of bark and bud or buds is applied to the matrix, tied with rubber bands or with cotton twine and sealed with melted paraffin.

At present, the principal use of the skin bud is in the budding of small nursery pecan trees. Doubtless it could be used upon hickory nut and walnut also. Skin budding may be done from early spring until late fall. Current-season buds may be used; or previous-season buds may be used from storage or direct from the tree as needed during the budding season.

Chip Budding.—Budding by this method may be done at seasons of the year when the bark does not slip. The usual season for chip budding is a period beginning about 2 weeks

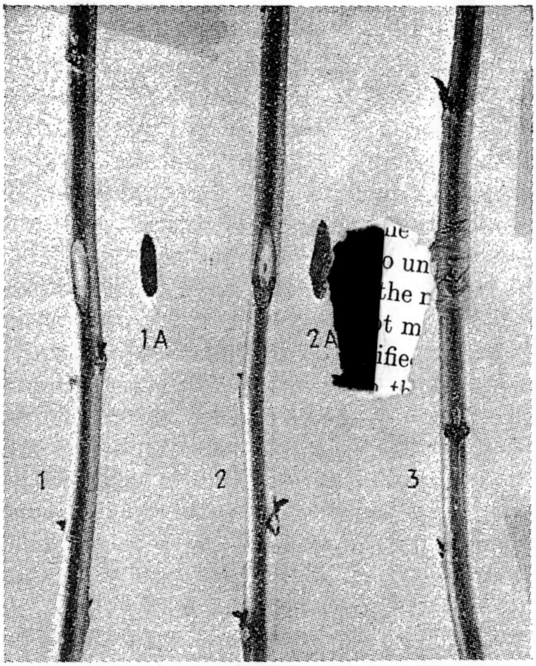

FIG. 132.—Steps in skin budding. (1) Stock; (1*A*) sliver of bark removed from stock; (2) bud stick, and (2*A*) bud removed from node; (3) bud applied to stock and tied with rubber budding strip.

before active growth starts in the springtime and continuing for about 5 or 6 weeks. Nurserymen sometimes resort to chip budding in the summer when drought causes stocks to become partially dormant. Under such conditions the T and patch bud cannot be used effectively, but the chip bud can be used provided the stocks are not too thoroughly dormant.

It is important to select a smooth place between nodes on the stock at which to insert the bud. A long downward cut through the bark and slightly into the wood of stock is first made. The cut should be 1 or 1¼ inches long and should constitute a smooth plane. A second cut is made downward at an angle of about 45 degrees so as to intersect the lower portion of the cut first made. As a result of these two cuts a chip of bark and wood is removed from the stock.

The bud from the budwood is removed in the same manner, beginning the first cut ¾ inch above the bud and the second one ⅓ or ½ inch below it. It seems to be important that the bud be

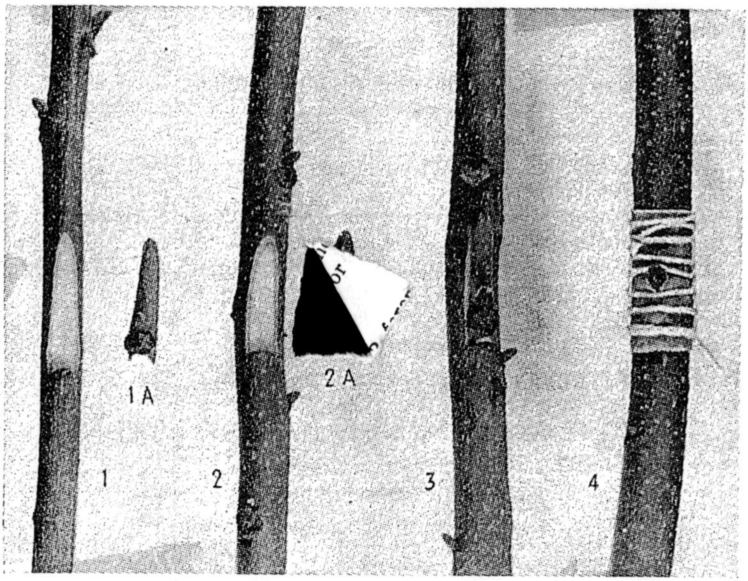

FIG. 133.—Showing different steps in inserting chip bud. (1) Scion; (1A) bud; (2) stock; (2A) chip removed from stock; (3) bud in place on stock; (4) bud protected with waxed cloth and tied with string.

on the lower and, consequently, thicker part of the chip. The chip is removed from the stock and the bud slipped into its place so as to make a close fit. Generally the bud will not fit perfectly on both sides; but it is important that the two cambium layers coincide on one side at least. Chip buds should be tied securely in place; if cotton string or raffia is used for tying them, wax, melted paraffin, or waxed patches should be applied to prevent drying. If waxed tape, rubber budding strips, or budding tape are used properly no other protective material is required for the bud.

For chip budding, the stock should not exceed 1 inch in diameter, and preferably should not be larger than ¾ inch. Budwood from which buds are secured should be thoroughly dormant at the time the budding is done.

Chip-budded trees have a long growing season and are ready for planting the fall following the spring during which they were budded. The system injures the stock but little and, if failures occur, some other method may be used during the same or a following season.

Questions

1. In what respects is budding different from grafting?

2. When is the favorable season of the year for T budding? Patch budding? Chip budding?

3. What is meant by dormant budding? June budding? Flipping the bud? Inverted T bud?

4. Under what circumstances is the patch bud used instead of the T bud?

5. What are the merits of the chip-bud method?

6. What kinds of budwood are used for T budding? Patch budding? Chip budding? Skin budding?

Suggested References

1. Bailey, J. E., and J. G. Woodroof: Propagation of Pecans, *Ga. Agr. Exp. Sta. Bull.* 172, 1932.
2. Hansen, C. J., and E. R. Eggers: Propagation of Fruit Plants, *Calif. Agr. Ext. Serv. Circ.* 96, 1936.
3. Lewis, I. P.: Grafting and Budding Fruit Trees, *Ohio Agr. Exp. Sta. Bull.* 510, 1932.
4. Reed, C. A.: Nut-Tree Propagation, *U. S. Dept. Agr., Farmers' Bull.* 1501, 1926.
5. Schrader, A. Lee, and W. E. Whitehouse: The Budding and Grafting of Fruit Trees, *Md. Agr. Exp. Sta. Bull.* 278, 1925.

CHAPTER XIV

PROPAGATION OF CERTAIN PLANTS

The general methods of propagation have been outlined in preceding chapters, but the specific practices followed in the propagation of different plants involve a more detailed consideration. Special attention is directed to the propagation of the more important fruits and of a few other representative species.

Peach.—The peach and other stone fruits are relatively easy to propagate by budding. Grafting is seldom used on these plants, either in the nursery or in top-working, because gum forms readily at the points where grafts are inserted and interferes with the union of the stock and scion. The gum may not prevent the union entirely in every case, but it will likely prevent a strong union. Budding is usually done on young wood, since gum does not form so readily on it.

1. *Rootstocks.*—For peaches the common commercial rootstock is the small seedling peach of the southeastern mountain areas, particularly Tennessee and the Carolinas, the seed of which are often designated as *naturals.*

It produces a good root system, makes a strong, vigorous growth, is adaptable to a range of soil conditions, and is congenial with most varieties of peaches. It is a satisfactory stock from these standpoints and is used more extensively than any other. This rootstock, unfortunately, is subject to the attack of nematodes, discussed in the following chapter, and that is the basis of serious objection to using it where nematodes are present.

The Marianna plum grown from hardwood cuttings is used to some extent in the South as a rootstock for peaches. It is reported to be highly immune to nematodes but apparently is not a suitable rootstock for some varieties.

The Shalil peach, introduced from India in 1913, is apparently immune to nematode damage and for this reason is attracting attention as a new rootstock for peaches. It does not become completely dormant in some climates, and that may prove to be objectionable. It will take many years to test it adequately and determine its enduring value as a rootstock. Another stock, introduced by the United States Department of Agriculture and known as F. P. I. 61302 is likewise promising because of its resistance to nematodes. The apricot is highly immune to nematode damage and is used to a limited extent as a rootstock for that reason, despite the fact that it has a dwarfing effect on peach tops. A Chinese wild peach (*Amygdalus davidiana*) definitely tolerant of alkali has been used satisfactorily as a rootstock where peaches are grown on soil containing excess quantities of alkali. Peach is also grown on rootstocks of Myrobalan plum, seedlings of named varieties of peaches, seedlings of cannery pits, and almonds.

Rootstocks for peaches are grown or produced principally from seed. Two important factors influence the germination of peach seed: the strong covering and afterripening. Soaking or stratifying the seed after it has once been allowed to dry tends to cause the bony covering to crack and become weakened so that the expanding embryo can split it open. Storing the seed for 8 or 10 weeks under moist conditions at a temperature slightly above freezing enables the seed to complete afterripening processes, which usually must be accomplished before the seed will germinate. The cold storage at a uniform low temperature is better than the alternating high and low temperatures under natural conditions. Some work with sulphuric acid has indicated that treatments of 5 to 15 minutes in 25 and 50 per cent acid are effective in hastening germination and increasing the percentage. The seed are treated just previous to planting and should be washed carefully after removal from acid. It is thought that the acid, in addition to increasing the permeability of the outer covering, also has some influence on the rest period.

The Marianna plum, when used as a stock for peaches, is commonly grown from hardwood cuttings. These are made from dormant shoots about the size of a lead pencil; they are cut to a length of 6 or 8 inches during the winter and held in cold storage at a temperature of from 32 to 36°F. until time for planting. Shortly before the beginning of growth in the spring they are planted in nursery rows with a spacing of 8 or 10 inches. They ordinarily root readily, make good growth, and are large enough to bud the following summer or fall.

2. *Budding.*—In the case of peaches the usual method of budding in commercial propagation is the T bud. Other methods may be used under special or unusual circumstances.

Budwood for budding the peach by the T-bud method is usually taken from shoots of the current season's growth. The buds are formed in the leaf axils and are satisfactory for budding work as early as May or June in the South, even though they

FIG. 134.—Peach seedlings ready to bud, in late summer.

may appear to be quite immature. In some cases the leaves are clipped off several days in advance so that the petioles will fall off before the buds are used. The common practice, however, is to cut the leaves off at the time the bud sticks are cut; a portion of the petiole is left and is useful in handling the bud when it is being set in the stock.

Where the growing season is long, peach pits planted in fall or winter germinate in early spring, and the young seedlings are large enough to be budded in June. The buds are forced at the time of budding, and the resulting trees, known as *June buds,* are ready for planting the following winter, a year from the time the seeds were planted. June-budded trees often grow from 2 to 3 feet between the time the buds are set in the summer and the end of the growing season in the fall; under very favorable conditions they may make 4 or 5 feet of growth.

In the North the seedling trees are seldom large enough to be budded before late summer. They are budded mainly in August; the buds remain dormant over winter and are forced the following spring. The buds start growing in the early spring and make large, well-developed trees by fall. They have a two-year rootstock and a one-year top, and are sold as one-year-old trees.

The commercial budding crew consists of one budder, who merely inserts the buds, and from one to three less skilled workers to tie the buds. The buds are usually cut from below, and left attached to the bud stick by a chip of wood, before the budder begins in the field. A number of bud sticks are prepared and wrapped in damp cloth. A helper may prepare others as needed. As the buds are to be inserted they are cut across the top of the shield, through the bark only, and popped loose from the chip of wood. A good budder may put in from 1,500 to 2,000 buds in a day and keep 2 or 3 men busy tying. Cotton twine, raffia, and corn shucks torn in strips are used for tying. Rubber bands are also used extensively; these will disintegrate in a few weeks, and the labor and expense of removing them is eliminated, as is also the danger of girdling the stock by leaving the tying material about the bud too long.

Drought or other unfavorable growing conditions sometimes cause the bark of the stock to "stick" or "set" during the budding season. In such cases, the chip-bud method may be used instead of the T bud. The chip bud may also be used in the early spring on one- or two-year-old stocks, in which case dormant budwood is required.

Peaches may be root-grafted with the whip graft. Stocks that have completed one season of growth are whip-grafted at the crown with dormant scions. This is done in late winter or very early spring. Root-grafted trees are more likely to develop crown gall, and the practice is objectionable for this reason.

3. *Top-working.*—It is frequently desirable to change the variety of large peach trees by top-working. This may be done

in several different ways. Probably the most satisfactory system is one by which the top is dehorned and the new shoots that develop afterward are budded to the desired variety. Several steps are involved.

a. Dehorning.—This should be done when the trees are dormant, preferably in late winter. The top should be cut back to points where the framework branches are from 1 to 2 inches in diameter. As far as possible limbs should be cut so as to result in a dehorned tree that is symmetrical in shape.

b. Thinning the Young Sprouts.—A large number of young sprouts will develop on each stub of the dehorned tree during the spring. These should be thinned out when the largest are 6 or 8 inches long. Four or five of the largest and most vigorous shoots are left. It is important that one shoot be allowed to grow near the end of each stub and that all those that are left be properly spaced along the stub; on stubs that extend out in a horizontal direction, only shoots that grow out from the upper side should be selected. In thinning the competing shoots, it is not always necessary to cut them off. They may be twisted and partially broken in such a way that they will continue to support the tree and prevent sunscald and will at the same time not interfere with the growth of the selected shoots.

c. Budding.—The shoots that begin to develop on stubs in early spring and are selected somewhat later may be budded in early summer and the buds forced into growth the same season; or the budding may be done late in the summer and the buds, which remain dormant over winter, forced into growth the following spring. Though all the selected shoots on a stub are usually budded, two successfully budded shoots on each stub are normally sufficient.

d. Aftercare.—It is necessary to recut the stub back to the nearest growing shoot if none happens to grow on the end of the stub. Native shoots may continue to develop on the old framework for several seasons. These should be cut out as often as necessary. When the budded top has grown sufficiently to provide shade for the framework, native sprouts do not develop so freely.

The cleft graft, bark graft, and inlay graft are methods that are occasionally used in top-working large peach trees. These methods may be used with a fair degree of success provided the grafts are set in limbs that are 1 inch in diameter, or less. If limbs much larger than this are grafted the formation of gum by the stock is likely to interfere with the union; this tends to restrict the growth of the graft and delays the healing of the wound by overwalling.

Plum.—The methods used in the propagation of the plum are generally identical with those used for the peach. The T bud is the most common method used; the chip bud and whip graft may be used in special cases. Trees to be top-worked are usually dehorned and budded at a later date on new growth. Plum trees, may, however, be cleft-grafted and bark-grafted more successfully than the peach.

For some sections, Myrobalan is the most satisfactory rootstock for plums. In New York, 15 varieties of plums, comprising European, Japanese, and several native species were tested for 20 years on four plum rootstocks and on peach. Results showed Myrobalan to be far superior to others tested. Apricot rootstock is used in sections where nematode injury is common. It makes a compatible union with plum and appears to give satisfactory results. The Marianna plum is used successfully in the South for stock. The Shalil peach, which is also nematode-resistant, is another possible stock for plums, though its congeniality with different plum varieties has not been tested adequately. In many cases nurserymen use identically the same rootstocks for plums as they use for peaches. This practice is satisfactory in some cases, but in general more discrimination is warranted since all varieties of plums do not respond in the same way on various rootstocks.

Apricot.—Seedling apricots, especially of Blenheim and Royal varieties, are the principal source of rootstocks for standard varieties of apricots. These form a congenial union and are resistant to nematodes. Tennessee or Carolina "natural" seedlings and seedlings from certain named varieties of peaches may also be used satisfactorily if the trees are to be grown in soils free from nematodes. Myrobalan plum is used in some cases where the apricots are grown in heavy soil. Apricot trees suitable for planting are usually propagated in the same manner as outlined for peaches, and trees to be top-worked are either (1) cleft- or bark-grafted or (2) dehorned and budded.

Cherry.—Two rootstocks have been used almost exclusively in the propagation of cherries. These are the Mahaleb (*Prunus mahaleb*) and Mazzard (*P. avium*), and there has been considerable controversy as to the relative merits of the two. It has been fairly well established that Mazzard is the more desirable stock from the standpoint of the orchardist. Trees live longer on this stock and withstand drought and diseases better than on Mahaleb. It is considered that the Mazzard rootstock is more subject to injury by winter cold than Mahaleb; this apparently is true of trees in the nursery and also for larger trees in the orchard. Seedling trees of Mazzard are more difficult to produce in the nursery, an account of leaf diseases. Mahaleb trees make better growth in the nursery and are larger and more uniform than Mazzard of the same age, but they grow more slowly and are less vigorous as they become older, and sooner or later they exert a dwarfing effect on the top. Briefly, Mahaleb makes better growth in the nursery, but Mazzard produces the better orchard trees. In recent years Morello stock has been used to good advantage in California in heavy, shallow, or wet soil. It has some dwarfing effect and does not make a good union with all varieties.

The germination of cherry seed is frequently a matter of some uncertainty. Many seeds, especially of sweet cherry, are not viable. The cherry pit absorbs water readily through the bony endocarp that surrounds the seed, but the covering may be more easily broken by the expanding embryo if it is weakened by alternate wetting and drying. Cherry seed do not germinate well unless the rest period is broken, and this is accomplished by stratification at a cool temperature.

Almond.—The bitter almond is used as a stock for the almond, but the sweet almond is also satisfactory. Almond stock will live in comparatively dry soils, and, although it may not support heavy crops under such conditions, it will carry the tree through very dry seasons and enable it to resume growth when conditions become favorable. Peach is sometimes used as a stock for almond, and the union is entirely satisfactory. Chinese peach likewise is a promising stock for almond where alkali may be troublesome. The Myrobalan plum has been used as stock, but almond tops overgrow it. Apricot is sometimes used but makes a poor union with almond.

Seedling trees to be used as rootstocks for nursery almond trees are usually T-budded. Old trees may be top-worked to other varieties by either (1) cleft grafting or (2) cutting the top away during the dormant season and budding on young sprouts the following summer.

Cherry Laurel.—This popular broad-leaved evergreen plant is closely related to the stone fruits. It may be propagated by means of cuttings of ripened wood, by root cuttings, and by layers. Plants for ornamental purposes are sometimes grown from seed. Named varieties may be grafted or budded on seedling stocks and have been grown experimentally on peach and plum.

Apple.—The apple is one of the most popular fruits in the United States and is perhaps grown more generally than any other. Propagation practices influence in several ways the performance of apple trees, and for this reason methods of propagation should be a matter of concern to the apple grower as well as to the nurseryman.

1. *Rootstocks.*—Many different stocks are used in the propagation of apple trees. The standard one for many years was the seedling French crab. In some cases the seedlings were grown in France and imported; in others seeds were imported and the young plants grown in America. More recently, the Virginia crab has been used and is considered to be a superior stock for certain varieties. Seedlings of several named varieties, notably Ben Davis, Winesap, McIntosh, Whitney, and Northern Spy, are satisfactory from the standpoint of vigor and uniformity. Clonal rootstocks of Hibernal, Northern Spy, and Delicious are used to adapt varieties to certain unfavorable conditions.

Cider pomace is the principal source of seed, whether Virginia crab, French crab, or some named variety is used. It is important that the seed be separated from the pomace while it is fresh and before it begins to heat as a result of fermentation. The seed are washed, stratified, and kept continuously moist. In addition to this, they should be subjected to a temperature of from 34 to 38°F. for 8 to 10 weeks in order to break the rest period. Both of these effects are accomplished by storing them in loosely filled bags between cakes of ice. Seed may be planted in flats or cold frames in the spring, and the seedlings transplanted to the open field as soon as the second pair of true leaves is formed. These small plants are transplanted with much greater success than are the larger ones. Seed may also be planted directly into rows in the nursery. In either case, the seedlings are commonly allowed to grow 1 year in the nursery and, in some cases, 2 years.

The quality of seedling rootstocks is determined largely by the soil in which they are grown and the treatment given them during the first year of growth. A deep, fertile, well-aerated soil encourages the growth of seedlings with long, straight, thick

FIG. 135.—One-year-old apple-seedling rootstocks and scions for whip grafting.

taproots that have few side roots. These are suitable for piece-root grafts, one root often providing 2 or 3 piece roots. Some propagators prefer whole roots with well-developed branch roots for rootstocks to be used in either budding or grafting. The formation and development of branch roots can be encouraged by severing the taproot of the young seedling 3 or 4 inches below ground. This may be done with a long, sharp knife, and irrigation should follow immediately in order to settle the soil around the roots. One-year-old seedling plants that are lined out in the spring for budding the following summer produce branched roots as a result of having been transplanted. Most of the apple-seedling rootstocks in the United States are grown in the Kaw River valley in Kansas and on the Pacific coast, particularly in Washington.

FIG. 136.—Two-year budded apple trees in nursery row. (*Courtesy A. F. Lake, The Shenandoah Nurseries, Shenandoah, Iowa.*)

Clonal stocks, which have received considerable attention in England, have been developed, in order to avoid the variability induced by seedling rootstocks. These clonal stocks have been produced by mound layerage and, in some cases, by cuttage. Results in England and in this country indicate that trees on such stocks show slightly less variability than others on seedling roots, but on the whole the difference has not been so great as anticipated.

2. *Budding.*—Two, and sometimes three, years are required to produce an apple tree with a one-year top of standard variety. Seedling rootstocks that have grown 1 year from seed are dug in the fall or winter, lined out in the nursery row, and budded during the second season. If the budding is done in early summer the bud may be forced to make sufficient growth the same season to be suitable for orchard planting the following winter. This practice is followed where the growing seasons are long. More often the one-year seedlings are lined out at beginning of the second growing season, and budded during the late summer. The buds remain dormant over winter, are forced the following spring, and make 4 or 5 feet of growth by fall. These are known as *yearling* trees. Trees of about the same size and grade can be produced in two seasons by planting the apple seed in pots in a greenhouse in late winter; the resulting seedlings are shifted to the field in early spring without disturbing the root system. These are large enough to bud by late summer of the first season, and the buds produce good top growth the second season if properly forced.

The T bud is the prevailing method of budding the apple. It has long been the practice to insert T buds in apple stocks slightly below the ground line. To do this it is necessary to clear the soil away and wipe the adhering soil from the stock. It has recently been shown that buds set 3 to 4 inches above the crown begin growth the following spring more evenly and make taller and larger trees than those inserted at a lower level.

Buds that are set in late summer and remain dormant over winter are forced the following spring by cutting the stock off 2 or 3 inches above the bud. As soon as the bud makes from 4 to 6 inches of growth, the stock is recut immediately above the growing shoot. Current-season buds are ordinarily used for budding apples, though one-year-old scions that have been kept dormant in cold storage may be used as a source of buds for early budding.

The chip-bud method may also be used successfully in budding apples. It is used in the early spring on one- or two-year-old rootstocks, and the buds if forced immediately will make good growth during the following season.

3. *Grafting.*—Seedling apple trees that have grown 1 year may be grafted in the nursery or bench grafted. In either case the whip graft is the method used, and scions are inserted on the root below the crown.

If apple trees are grafted in the nursery, the work should be done in late winter or very early spring when temperatures are likely to be favorable. Soil is cleared away and grafts are placed on the root. The union is wrapped with cotton string or tape and moist soil is pressed about the scion and union, leaving only the top bud exposed. It is not necessary to remove the cotton string since it soon decays; the tape must be removed within 1 month or 6 weeks in order to prevent girdling.

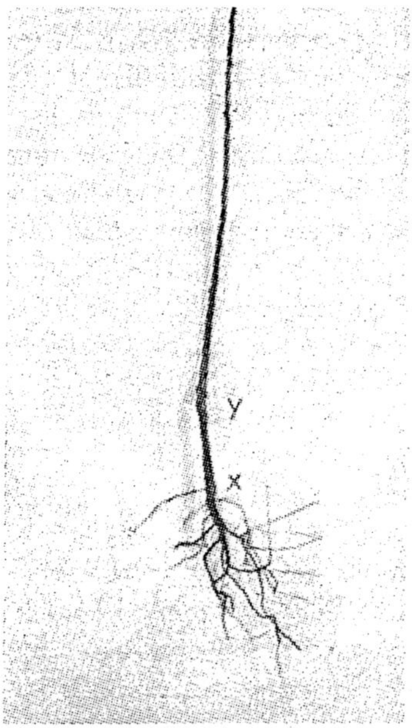

FIG. 137.—An apple tree as it appeared one year after the scion was grafted on the seedling rootstock at *X*. The top of the scion at the time it was grafted on the stock is shown at *Y*.

The more common practice is the bench graft. The seedlings are dug in the fall and grafted indoors during the winter months. As soon as the grafts are made and wrapped they are packed in moist, well-aerated material and allowed to callus for a period of 10 days to 2 weeks. A temperature of from 75 to 80°F. is favorable for this. Callused grafts may be planted out in the nursery row immediately; or if planting is delayed by unfavorable weather conditions, they may be held for several weeks in cold storage at a temperature of about 32°F. In planting apple grafts in the nursery, they are spaced 8 or 10 inches apart in rows that are 5 or 6 feet apart. It is important that individual grafts be placed so that the union is slightly below the ground level. The most desirable shape of tree is produced if only one shoot is allowed to grow from the scion; others should be pruned off at intervals during the first season. Such grafts may make from 4 to 6 feet of growth during the following season and are large enough for orchard planting the following fall or winter. For bench grafting, propagators prefer seedlings with long, straight taproots and few large side branches.

Own-rooted or *scion-rooted* trees are produced by grafting a long scion onto a relatively short nurse-rootstock. After the two have callused, the graft is planted so that the union is considerably below the ground, and soil is pressed about the scion. Roots sometimes develop on the lower portion of the scion during the first season. *Scion shoots* of some varieties that are caused to grow from below ground by deep planting are more likely to form roots than the old scion. Inverted nurse roots are also used in encouraging the rooting of apple scions. The chief value of own-rooted apple trees is for use in experimental work.

Double working is frequently used in propagating apple trees to discourage the development of crooked or scraggly growing trees and to prevent damage from collar rot. The Grimes Golden variety is quite susceptible to collar rot, while the Delicious is relatively immune. Delicious scions, when used as splices at the ground line between the seedling rootstock and the Grimes Golden top, produce trees that are seldom troubled by the disease.

Apple trees that have been propagated by whip grafting onto a root used as a stock frequently develop callus knots, crown gall, hairy root, and similar kinds of abnormal growths. The union of the stock and scion often results in callus overgrowths that are physical abnormalities. In other cases the succulent callus tissue presents a favorable medium for the entrance of organisms that cause crown gall and hairy root. Disinfecting stocks and scions before preparing them, and also after they are joined together, by dipping in Bordeaux mixture or bichloride of mercury solution is effective in reducing crown gall and hairy root. It is well also for the propagator to sterilize his knife and hands at intervals. Callus overgrowth can be prevented to some extent by wrapping the grafts with waxed tape or medicated tape instead of string. The tape will girdle the stock if it remains too long, but in the meantime it will usually have caused the stock and scion to make a smooth union. Despite these precautions abnormal overgrowths on whip-grafted trees are of frequent occurrence in some seasons.

4. *Top-working.*—Large apple trees may be readily top-worked by either budding or grafting. The more common practice is to dehorn the old tree by cutting limbs off at points where they are from 1 to 3 inches in diameter, and to insert one or two cleft grafts in each stub. The apple is relatively easy to graft, and cleft grafts unite readily if the work is done with reasonable care. The bark or inlay graft can also be used successfully on limbs that have been cut back as for cleft grafting. By either of these methods the new top of a standard variety makes a vigorous growth and may replace the old top in one or two seasons.

By another practice the tree is dehorned in the winter, the young sprouts that develop on the stubs are thinned in early spring, and the selected ones are budded when they attain sufficient size in the summer. If the shoots become large enough to be budded in early summer the buds are forced out immediately. Buds that are set late remain dormant over winter and are forced the following spring.

5. *Dwarf Apple Trees.*—Apple trees may be grown as *dwarf* trees by the proper selection and use of a rootstock and by giving the tree the necessary aftercare. Paradise is a French stock that causes the top to be definitely dwarfed, and the Doucin exerts a partial dwarfing effect. These two rootstocks are ordinarily grown by mound layerage. Dwarf apple trees are more compact than standard trees of the same variety, and, although they bear high quality fruit, the trees remain permanently dwarfed in stature. They are quite popular in European orchards and were frequently planted in the older orchards in this country but are seldom seen now. They are interesting novelties but of little commercial value.

Pear.—The methods used in propagating the pear are almost identical with those used on the apple. It apparently forms callus freely and responds readily when either budded or grafted. The standard procedure in producing a tree to be planted in the orchard is to grow a seedling for rootstock and bud or graft a top of standard variety upon it.

1. *Rootstocks.*—Most pear rootstocks are grown from seed. Pear seed is usually somewhat more expensive than apple seed because of the greater difficulty involved in extracting them from the carpels of the fruit. The seeds require special treatment

in order to obtain good germination. It is important to keep them under moist conditions continuously from the time they are collected until they are planted; in addition they should be stored at a temperature of from 32 to 38°F., which serves to break the rest period and to prevent premature germination after the rest period is broken. The combination of moist conditions and a low temperature may be conveniently provided by placing the seeds in loosely filled bags and storing them between cakes of ice. Seeds are commonly planted directly into the nursery rows, and the resulting seedlings are ready to be budded or grafted during the second growing season. They may or may not be transplanted to a new location at the beginning of the second growing season.

Several different kinds of rootstocks are used for pears; they can be classified in three groups. (*a*) Until recently the French pear (*Pyrus communis*) was the only rootstock used for pears in the United States; it is still used more extensively than any other. It is well adapted to a wide range of soils and climatic conditions and is resistant to mushroom root rot. Unfortunately, it is quite susceptible to pear blight, though some varieties, such as the Old Home and Variolosa are quite resistant to the disease. French seedlings are also highly susceptible to the pear woolly aphis, which is a troublesome pest of the pear. The Bartlett, Winter Nelis, Anjou, and Hardy are French varieties that produce seedlings of desirable type. These varieties are grown in America and seed is generally available. (*b*) Certain oriental pears, notably *P. serotina*, *P. ussuriensis*, and *P. calleryana*, have been used as rootstocks for standard varieties. The so-called "Japanese pear" (*P. serotina*) is relatively immune to pear blight and woolly aphis. It does not thrive on wet soils and is not so resistant to mushroom root rot as French pear stock. Trees of this species have undesirable habits of growth; most of them have heavy thorns and few possess a desirable framework for top-working. Certain French varieties when top-worked on this stock produce fruit affected with a physiological disease known as *black end*. *Pyrus ussuriensis* has many characteristics that recommend it as a rootstock, but it is not recommended as a body stock, since some of its seedlings are susceptible to pear blight and many of them develop a framework that is not especially suitable for top-working. The

Callery pear (*P. calleryana*) is definitely promising as a rootstock for pears. Tops of standard varieties form a good union with this species, and make vigorous and uniform growth. Seedling trees of it are resistant to blight and woolly aphis and endure drought remarkably well. (*c*) Hybrid pears, such as the Kieffer, produce good seedling rootstocks, but they are not so resistant to fire blight as the pure oriental species and are no better than the French pears in other respects.

2. *Budding and Grafting.*—Seedling trees that have grown one season may be (*a*) dug and bench-grafted, using the whip-graft method, during the first dormant season; (*b*) whip-grafted in the nursery in late winter preceding the second growing season; or (*c*) budded during the second growing season either in their original location or after having been dug and lined out in a new place.

One-year-old pear seedlings that are whip-grafted, either indoors or in the nursery, usually make strong and stocky trees during the following growing season. Those that are grafted indoors are usually stored for 1 week or 10 days in moist insulating material at from 75 to 80°F. in order to encourage the formation of callus before they are lined out in the field. If budding is practiced the seedling trees are budded during their second year of growth. Where growing seasons are long, T buds, and rarely chip buds, may be set in early spring, and the budded tops if properly forced may make as much as 4 or 5 feet of growth during the first season. Where growing seasons are relatively short, buds are usually inserted in the summer of the second year but not forced until the following spring. With proper aftercare the following year the budded tops make strong and stocky trees suitable for orchard planting.

Crown gall, hairy root, mushroom root rot, fire blight, woolly aphis, and nematodes are hazards to be considered in propagating the pear. Close attention to the choice of rootstocks and methods of handling will aid in reducing losses from these agencies.

3. *Top-working.*—Pear trees may be easily top-worked by (*a*) cleft grafting, (*b*) bark or inlay grafting, and (*c*) dehorning during the dormant season and budding on young sprouts the following summer. Details of these operations and aftercare are practically the same as outlined for the apple.

Top-working is practiced much more extensively on pear trees than on most other fruits. The practice is widely followed in (*a*) changing trees of poor varieties to more desirable kinds and in (*b*) producing trees with a framework that is resistant to pear blight. Certain oriental species, such as *Pyrus serotina*, *P. ussuriensis*, and *P. Calleryana*, are definitely resistant to pear blight. According to one practice a tree of one of these species is allowed to grow until it forms a trunk and frame-

FIG. 138.—Japanese pear body stock top-worked with Kieffer pear at points indicated by an *X*.

work, and it is then budded or grafted with the desired variety. In some sections the resistant trees are set in permanent places in the orchard and trained to conventional form. When they have grown 2 or 3 years the scaffold branches are top-worked to standard varieties. Many oriental seedlings are unsuited for top-working due to their heavy thorns, poor shape, and variability in resistance to blight. Oftentimes the top-working operation is staggered over a period of several years; the lower limbs are budded or grafted at first, and the upper limbs are changed over in succeeding years, after they have made considerable growth. The resistance of the budded or grafted top of such trees to pear blight is determined by the variety used, but the trunk and main branches are relatively immune to damage. If climatic conditions during a certain season are especially favorable for blight it may kill the top back to where it was top-worked onto the stock, but the tree will be relatively easy to recondition by virtue of having the healthy framework.

The black end disease oftentimes becomes troublesome when some French pear varieties are budded or grafted on certain oriental rootstocks. It is found commonly on *Pyrus serotina* stocks and is also associated with *P. ussuriensis* and Kieffer. The Old Home and Surprise are two stocks that are resistant to fire blight and that apparently do not cause black end, and their use is being encouraged. Double working is frequently practiced in connection with the use of the latter two varieties and also with the Variolosa. Scions of these are whip-grafted on French seedling roots and allowed to grow until they develop a trunk and framework of desirable size. They are then top-worked to Bartlett, Seckel, Bosc, or other choice varieties. This procedure makes it possible to provide the same degree of resistance to blight as where the oriental stocks are used, without incurring the serious consequences of black end of pears.

4. *Dwarf Pear Trees.*—Certain pear varieties are oftentimes grafted or budded on quince stock. The two do not form a perfect union, and the resulting pear tree is dwarfed. Some varieties do not unite well with quince, and others such as the Kieffer make a strong growth and become top-heavy. The Bartlett forms a satisfactory dwarf tree on quince. Varieties that do not unite well may be double-worked on a Bartlett splice. Dwarf trees are more resistant to fire blight by virtue of making a less succulent growth than standard trees.

Quince and Related Plants.—Horticultural varieties of quince may be propagated by budding or grafting on quince stock. They may also be grown from cuttings and mound layers. Quince stocks for grafting may be grown from seed, stem cuttings, or layers.

Japanese *flowering quince* can be grown from cuttings, by graftage on quince stock, or by separation of suckers from a parent plant. Standard or tree forms have been produced experimentally by budding or grafting on apple and pear stocks. A distinguishing characteristic of such plants is that they do not produce suckers.

FIG. 139.—One-year-old pear top on quince rootstock.

Photinia (*Photinia serrulata*), a valuable evergreen ornamental plant, is commonly budded on quince stocks. The trees are dormant-budded in late summer, and the buds forced the following spring. The *loquat* (*Eriobotrya japonica*), a subtropical plant valued as an ornamental and also for its edible fruit, may be budded on loquat seedlings or on quince stocks. These plants to be used for ornamental purposes are frequently grown from seed.

Pecan.—A number of stocks are used for propagating the pecan, the most popular species being the common pecan, *Hicoria pecan*. Other stocks may be used. Native hickory trees (*H. alba*) are quite commonly top-worked to pecan. The two unite fairly readily, and their growth and fruitfulness seem to be determined largely by the soil in which the trees grow. Hickory nut trees in coarse sandy soil do not support a pecan top adequately, but in deep alluvial soil this difficulty is overcome. The bitter pecan, or pignut (*H. glabra*), which grows commonly in low swampy places, is frequently top-worked to pecans and apparently is a congenial stock. It is considered that this stock withstands transplanting better than the common pecan, though experimental evidence is lacking. Pecan stocks are grown only from seed, which may be planted in nursery rows in the fall, stratified over winter, and planted in the spring, or planted directly in the nursery in early spring after having been soaked in water or chemicals before planting. Pecan seed frequently sprout in the hulls before harvest, and they will germinate at weekly intervals throughout the winter. These phenomena indicate that the rest period is not a limiting factor in germination. They lose their viability if they become rancid, and, in order to prevent this, it is customary to keep planting seed in cold storage, especially if they are to be kept 1 year or longer before planting. Standard varieties cannot be reproduced true to type from seed, but certain selected seedlings are known to produce stocks of exceptional vigor.

The patch bud and bark and inlay grafts are methods of graftage used commonly in top-working pecan trees. The skin bud, and other modifications of the patch bud, and the whip graft are sometimes used. Trees that are 1 inch or less in diameter are relatively easy to bud or graft; probably the patch bud is more popular for this size than any other method. Whether this or one of the other acceptable methods of graftage is used it is advisable that the work be done quite early in the spring, and that the tree be forced heavily at the time the work is done.

Frequently it is desirable to top-work trees that are 6 or 8 inches in diameter. This may be accomplished by using either the patch bud or bark or inlay graft. Patch budding in which the buds are inserted in thick rough bark that must be pared down to the thickness of the bark of the patch bud is known as *direct budding.* Buds are inserted in limbs that are 3 inches or less in diameter. When this system is followed, the radical dehorning process practiced on larger trees is avoided. Skilled top-workers become adept in the art of placing the buds so that they will grow into a top of good conformation. It is not necessary to bud all the limbs of a tree; a minimum number of buds

FIG. 140.—Large top-worked pecan tree, showing budded top that has grown one season.

usually make as much total growth as a much larger number. The bark graft might also be used in top-working trees of this size. As a rule, however, propagators get a higher percentage of patch buds to make successful union than of bark grafts. Inlay grafts can be used quite effectively in top-working trees of this size, and they generally make a stronger union than bark grafts.

The system commonly employed in top-working large pecan trees involves dehorning the main branches. The dehorning should be done during the dormant season. Young shoots will develop on the stubs. These should be thinned out to 5 or 6 per stub early in the summer in order to encourage vigorous growth. The ones that are left may be patch-budded in late July or August following. In some cases the buds are forced at the time the budding is done; in others, they remain dormant until the following spring. The budding may be delayed until the following spring or a later one and the shoots handled much the same as nursery stocks. The patch bud or some modification of it is the favorite method for such stocks.

Walnuts.—Named varieties of black walnuts, English walnuts, and Japanese walnuts may be propagated by budding or grafting onto seedling stocks. The patch bud and bark and inlay grafts are the principal methods used. Some difficulty is encountered in budding black walnuts because of the tendency of the primary bud to abscise. The secondary bud changes into a catkin; the third bud of the group at a node is often difficult to force on account of its small size. This trouble is not met in grafting.

Walnut stocks are grown easily from seed. Black walnut (*Juglans nigra*) and Texas walnut (*J. rupestris*) are used commonly as stocks. Both of these stocks make slow growth in the nursery. The northern California black walnut (*J. hindsi*) is the most popular stock for English walnut. It makes exceptionally vigorous growth and is resistant to the oak root fungus but is susceptible to crown rot.

Filberts.—Several different methods may be followed in the propagation of filberts. Occasionally a tree grown from seed of a standard variety may have some merit, but the chances are it will be inferior to the parent tree. The filbert produces suckers freely from about the base of the plant. If soil is mounded up around the base of these shoots, they will develop a good root system in time and can be successfully transplanted. Filberts may also be propagated from cuttings. If the cuttings are taken from portions of suckers below the ground, they may root with a measure of success; those that are taken from above-ground parts root very poorly.

Filberts may be readily propagated by continuous layerage. Suckers from around the base are bent down to the ground and pegged in place in early spring. New shoots begin to develop from lateral buds, and as these grow they are covered by degrees until ultimately 6 to 8 inches of soil is added. These shoots will normally develop a good root system within one season and may be separated for orchard planting during the early winter. Where only a few plants are needed they may be grown from suckers that are tip-layered. Layered shoots from suckers produce plants of the variety represented by the root system of the original plant and do not reproduce varieties represented by the budded or grafted top.

Budding and grafting are sometimes used in propagating named varieties of filberts. Seedlings of commercial varieties and of the Turkish filbert are grown and whip-grafted to standard varieties. They may also be budded, the T bud being the method used. Large trees may be top-worked by using the cleft graft.

Tung Nut.—Trees used in the first commercial tung plantings consisted largely of seedling stock. The variability of seedling plants can be avoided by budding or grafting standard varieties on seedling rootstocks. The patch and ring bud are the two methods that are used most commonly. Seedling trees from seeds planted in February are budded during late summer or early fall. The inserted buds unite but remain dormant over winter. They begin growth the following spring, when they are forced by lopping the tops 5 or 6 inches above the bud. Blind buds are rather common on tung wood, and wood for budding should be selected with some care in order to avoid them.

Persimmon.—Both the native persimmon and the Japanese persimmon are regarded as difficult to propagate. Budding and grafting are practiced, but the percentage of success is often very unsatisfactory with either method.

1. *Stocks.*—In the Southern states the native persimmon (*Diospyros virginiana*) is used as a stock for both native and Japanese varieties. The stock is difficult to grow from seeds and is short-lived in many parts of the South. The Texas persimmon (*D. texana*) has been found undesirable as a stock for either type.

Seedling stocks of kaki persimmon (*D. kaki*) have given good results in California. More success has been obtained, however, from the lotus persimmon (*D. lotus*), which was introduced from China. It is almost the universal stock in that country and in

California, but neither this species nor the kaki is successful as stock in the Southern states.

2. *Budding.*—Very poor results are obtained in budding persimmons. In a few cases, T buds have been used with success in small stock, and the inverted T bud has given remarkable success in recent work. The patch bud, as used in pecan propagation, has been a failure in many cases. Ring buds are reported as successful in China in budding *Diospyros kaki* on *D. lotus.*

3. *Grafting.*—In the nursery, small stocks are whip-grafted to produce most of the commercial supply of trees. Trees of larger size are cleft-grafted while still in the nursery, where the bark graft may also be employed. Top-working by one of these methods of grafting is also practiced, but the results have been far from uniform.

Grape.—The grape is an easy plant to propagate. In commercial practice it can be reproduced from seed, from cuttings, by layering, and by grafting.

1. *Seed.*—Seeds taken from grapes at harvest time may be planted in the fall and allowed to remain in the soil over winter. They germinate the following spring when climatic conditions become favorable. More commonly they are stratified in boxes of sand over winter and planted early the following spring. A spacing of from 10 to 15 inches in the nursery rows is desirable.

Grape seedlings may be grown for two uses: (*a*) for grafting stock and (*b*) as a source of new varieties. From a group of seedlings there is the ever-present possibility that one of outstanding qualities will be found. Such popular varieties as the Concord, Carman, Campbell Early, Brilliant, Bailey, Moore's Early, and a long list of others have originated as seedlings. Because of the extreme variability of seedling plants of the grape, commercial viticulturists never grow a vineyard from seed.

2. *Cuttings.*—Most species of grapes can be grown from cuttings. Some varieties root more readily than others, and certain ones produce better root systems than others. Cuttings for propagation may be taken at any time the plants are dormant. They are ordinarily made from 10 to 15 inches long, the lower end being cut below a bud and the upper end 1 inch above the bud. Canes that are one season old, medium-sized, well-matured, and have short internodes are preferred. If the cuttings are made during the winter they may be held in cold storage until planting

time in the spring. Cuttings respond readily if planted in a well-drained sandy loam soil. They are planted in rows 4 or 5 feet apart and given a spacing of from 6 to 10 inches in the row. Individual cuttings are planted deep, leaving only one bud above the surface; to do this it is often necessary to set the cutting in a slanting position. Grape cuttings are used (*a*) as a means of

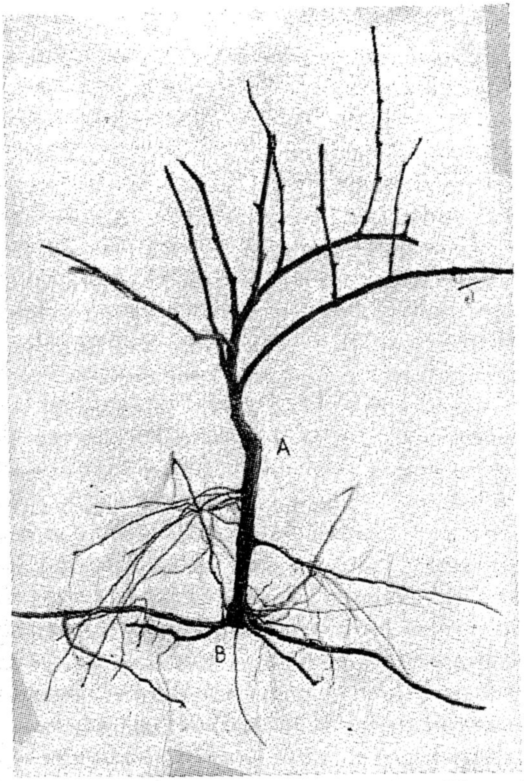

FIG. 141.—One-year-old grape plant grown from a cutting. The section between *A* and *B* represents the original cutting.

propagating new plants of certain varieties for planting in the vineyard without being budded or grafted and (*b*) to provide rootstocks that may be top-worked to standard varieties. Normally cuttings will make sufficient growth in one season to be used as vines for planting in the vineyard; they may, however, be allowed to grow two seasons in the nursery before they are transplanted. Cuttings that are to produce rootstocks for budded or grafted plants should be disbudded before they are planted. This is done by cutting off all except the bud nearest the top of each cutting. The treatment discourages the formation of shoots from below the graft union. Plants that are grown from cuttings for rootstocks are usually grafted during the winter following the first season of growth.

3. *Layers.*—Certain species of grapes do not root well from cuttings. This is true of the Muscadine grape (*Vitis rotundifolia*). Since they are also difficult to graft, layers are used to propagate named varieties. Simple, compound, and continuous layers are used under varying conditions.

4. *Grafting.*—There are several considerations that may prompt the grower to use grafted plants, instead of plants grown from cuttings, in establishing the vineyard. It is known that the cuttings of some varieties form poor root systems, and as a result the vines are not so productive as when grafted on some other rootstock. This seems to be true of the Delaware and Campbell, at least under some conditions. The root-knot nematode is one of the destructive insect pests of grapes. Workers of the United States Department of Agriculture found that 154 varieties of European wine grapes (*Vitis vinifera*) were very susceptible to root-knot infestation. The majority of the American species and hybrids tested were likewise found to be seriously affected when planted in nematode-infested soil. Some varieties tested have some inherent resistance to nematode injury, and this suggests their possible value for stocks. The grape root louse or phylloxera (*Phylloxera vitifoliae*) is also a troublesome insect pest on all wine grape varieties and on some American varieties. Varieties of some American species show marked immunity to phylloxera damage and are hence valuable as grafting stock for susceptible varieties. Root rot is a disease that attacks only the root system of grapes. Some species are resistant to it, others highly susceptible. These considerations and others suggest the possible value of graftage in successful culture of grapes in the different environments in which grapes are grown.

The whip graft is used almost exclusively in grafting small grape vines. One-year-old stocks grown from cuttings or seed may be whip-grafted in the nursery in late winter or very early spring preceding the beginning of the second year. These will produce strong, stocky plants during one growing season and will ordinarily be large enough to be planted in the vineyard the following winter or spring. Instead of grafting the stocks in the nursery they may be bench-grafted during the winter and replanted in the nursery in the early spring. If this procedure

FIG. 142.—Grape plant after one season of growth, produced by grafting a scion on an unrooted cutting. The point of union is shown at *A*, and the terminal end of the scion at the time it was grafted on the stock is shown at *B*.

is followed it is advisable to store the grafts until they form callus before planting them outside. Plants handled in this manner will likewise produce good vines in one season after having been grafted.

Quite commonly unrooted cuttings about 10 inches long are used for stocks. They are cut so as to have a node at the bottom and sufficient internode at the top to permit the slanting cut necessary to make the whip graft. The scion is made by cutting ½ inch above a bud and leaving sufficient smooth wood below to make the slanting cut. These cutting grafts are usually made indoors near the end of the winter season. The finished grafts are commonly tied into bundles and packed in moist sphagnum moss, sawdust, or other insulating material and held at a temperature of from 75 to 80°F. for 10 days or 2 weeks. During this time they form the callus that is essential to a union. Grafts that have callused may be held in cold storage until a convenient time for planting in the nursery. The planting, however, should not be delayed past the usual season when grapes begin growth. Some propagators follow a practice of planting the grafts immediately after they are made, even though it may be several weeks before the growing season. This practice is open to the criticism that the temperature in the nursery may not be favorable for callusing.

In lining out the grafts they are placed in a deep furrow with the unions at the level of the ground. A straightedged board is used to get them all at the same level. They are placed 8 inches apart in the furrow and the soil pulled in around them. As soon as possible the earth is mounded along the row so as to cover lightly the top of the graft. It is important that the soil be kept continuously moist around the graft union and scion, by irrigation if necessary. When the new shoots have grown through the soil, and the cutting stocks are apparently well established, the soil is removed to the level of the union, and all roots originating from the scion are cut off. Any shoots arising from the stock are likewise removed. Roots seldom develop on the scion after the soil is removed down below or to the union.

Old established vines are frequently top-worked by use of the cleft-graft method. They are cut off at a smooth place near the surface of the ground. The grafts are inserted and moist soil is pressed about them in such a way as to leave only the top bud of the scion exposed. For stock over 1 inch in diameter, two grafts are used, and both are allowed to grow if they unite. This may be done in late winter before the period when sap flow is excessive or for 2 or 3 weeks after that period, but grafts seldom grow if the grafting is done during the period when the vines will "bleed."

5. *Budding.*—Small grape rootstocks may be top-worked by budding instead of grafting. One-year-old plants grown from

cuttings are usually used for budding stock. These are dug up and prepared for planting in a temporary place or in the vineyard. The preparation consists of cutting off some side roots, leaving only those that arise near the base, and removing all buds from the original cutting to discourage the growth of suckers. Lateral shoots near the top of the cutting should be pruned back to two buds each. This rootstock is then planted so that 2 to 3 inches of the old cutting is above the soil level. Soil, however, should be mounded about the base of the cutting in order to protect it from the sun until it is budded.

The chip bud is the method commonly used when grapes are budded. Midsummer is the proper time for this practice. The buds are taken from budwood of the current season's growth, which has reached a stage of maturity that is indicated by the brown color of the wood and buds. The buds are inserted slightly below the ground level on the part of rootstock represented by the original cutting. They are wrapped with rubber budding strips, covered with soil to a depth of about 6 inches, and left without further treatment until the following season. In the meantime the bud is expected to form a union with the stock. When the vines start growth the following spring the mound of soil is removed, and the stock is cut off 1 to 2 inches above the chip bud; any roots that may have formed from the bud are trimmed off and the tying strips are cut also. After this the bud is again covered 1 or 2 inches with loose soil. When the shoots from the buds have made 8 or 10 inches of growth the soil is removed finally from about the bud union and all side roots are pruned off. Plants propagated in this manner ordinarily come into normal bearing by the fourth year.

6. *Stocks.*—The grape phylloxera, nematode, and root rot are pests that must be considered in selecting a stock for grapes. In addition, stocks respond differently in different kinds of soils. Stocks with pronounced resistance to phylloxera have been selected from *Vitis rupestris*, a common one being the so-called "Rupestris St. George." This stock has soil limitations, however, thriving in a deep permeable soil, but not in shallow, poorly drained, or clay soils. Many other stocks are known to be resistant to phylloxera, but their use is likewise complicated by soil limitations, difficulty in rooting cuttings, and difficulty in grafting. Nematodes attack most varieties of Vinifera grapes and the

majority of American species. Varieties of certain species appear to have some inherent resistance to nematode injury. Most promising stocks tested are the Dog Ridge, Barnes, De Grassett (*V. champini*), and Doan (*V. doaniana*). Resistance to phylloxera is no indication of resistance to nematodes; the phylloxera-resistant Rupestris St. George is highly susceptible to nematode injury. The Dog Ridge, Champanel, and La Pryor are three stocks recommended as being resistant to root rot.

The Bramble Fruits.—In general the bramble fruits are given relatively close spacing in the field. Consequently, large numbers of plants are required for an acre, and it is important that they be produced economically. Fortunately the different kinds of brambles are easy to propagate—by one method or another.

1. *Blackberry and Dewberry.*—Suckers from old plants are a source of new blackberry and dewberry plants. In using these, however, there is always the possibility of getting seedling plants mixed with them, and seedlings are not likely to be desirable plants. Standard varieties may be readily propagated on a commercial scale from root cuttings. Roots the size of a lead pencil or larger are suitable. These are usually planted in nursery rows in late winter and given a spacing of 6 or 8 inches. They may be cut about 2 inches long, dropped into a furrow, and covered with 3 or 4 inches of soil. They may also be cut about 5 inches long and planted in a vertical position with only the tip of the cutting exposed, in which case it is important that the portion of the root that was nearest the base of the mother plant be placed up. In making the cuttings the top end may be designated by cutting it squarely across, and the bottom end by cutting it at an angle. Root cuttings grow readily and produce large, strong plants in one season.

Blackberry and dewberry plants may be grown also from layers. Vines are covered their entire length in late winter or early spring. New shoots develop from lateral buds, and these form roots on the underground portion. The layers may be separated and transplanted late in the same spring in which they were started, or they may be allowed to grow an entire season before being moved. The Young variety produces natural layers freely, and these are a source of new plants. This variety may be grown also from softwood stem cuttings, which root fairly easily in sand.

2. *Raspberry.*—Red raspberries produce suckers, or shoots, which arise in great numbers from the root system after cultivation has ceased. Those that reach a height of 8 or 10 inches make excellent plants for starting a new plantation. They should be lifted with a spade with as much of the root system as possible and replanted in the new location.

Black raspberries and most purple varieties are propagated by tip layers. In many cases, tips of long canes are accidentally covered with soil and natural layers are formed. When many plants are needed, the necessary number of tips should be covered in late summer. With favorable soil moisture, a well-rooted plant will be formed by the end of the growing season. At planting time the following spring, the parent lateral is cut off a short distance above the ground and the new plant dug out with a good portion of the root system.

Strawberries.—The strawberry plant produces runners freely, which form "rosettes" and take root usually at every second node. These may be dug carefully and separated from the parent plant and used for starting a new plantation. Runners of the current season's growth, transplanted to the new location in late fall or early winter, make the best plants. The ground should be thoroughly prepared and the rosettes planted out so that the crown will be level with the surface of the soil. Those that are set too deep or too shallow do not make good growth.

Everbearing strawberries produce few runners, since the buds that normally grow into them form flower buds instead. This characteristic makes it necessary to propagate everbearing varieties by other means than rosettes. Fortunately, plants of such varieties develop several crowns by the end of the growing season. These plants may be divided by breaking or cutting the crowns apart. Only the larger crowns and those that have 10 or 15 roots are satisfactory for planting.

The Bush Fruits.—The blueberry, gooseberry, currant, and cranberry are the important bush fruits grown in the United States. They are spaced close together in a permanent planting, and relatively large numbers are required for an area.

1. *Blueberry.*—Some difficulty is experienced in propagating the blueberry, and this has retarded the cultivation and improvement of this native fruit. Investigations have shown that it can be grown readily from hardwood cuttings, that the method is

entirely practicable, and that plants can be produced at reasonable cost. The solar frame was first used successfully, but the box frame, which is easier to construct and cheaper, is just as satisfactory. The cuttings are made about 4 inches long from shoots produced the previous season, and planted out in late winter or early spring. German peat apparently is a better rooting medium than sand, sand and peat, or American peat.

2. *Gooseberry.*—This fruit is easily propagated by layers and cuttings. Common varieties are ordinarily started from mound layers. Stock plants from which layers are to be secured should be pruned back heavily before they begin growth in the spring. This stimulates the growth of several shoots from the base of the plant. In midsummer, earth is mounded about the plant so as to cover the lower portion of the new shoots to a depth of 4 or 5 inches. By fall the shoots will have rooted. During the winter the mound is removed, and those cuttings with well-developed roots are cut off and set in a nursery row where they are grown for 1 or 2 years before being finally planted in the field.

Gooseberries may also be grown from cuttings; some varieties root readily and others very poorly. Cuttings about 6 or 8 inches long are made from vigorous-growing shoots during the fall or winter. These are either stored in a cool place for spring planting or set directly into nursery rows. They are spaced from 4 to 5 inches apart and set at a depth that permits only two buds to extend above the ground. New plants suitable for field planting are produced from cuttings in one or two seasons.

3. *Currant.*—New currant plants are grown almost entirely from cuttings made from vigorous shoots that have completed one season of growth. They are cut about 8 inches long and stored in a cool place during the fall or winter for spring planting, or planted as soon as they are made. In either case they are set in rows with from 4- to 6-inch intervals between plants and at a depth that permits only one or two buds to extend above the ground. Better rooting is obtained if the soil in which the cuttings are placed is well-drained and is sufficiently loose and porous to permit good aeration. Some cuttings will produce plants suitable for field planting in one season; others require two seasons.

4. *Cranberry.*—The cranberry is propagated commercially from cuttings, which are planted directly in place without previous rooting. Most cranberry plantings are made on acid peat soils.

A surface layer of sand, 4 or 5 inches deep, is added to the peat before the area is planted. The cuttings are pressed into the sand so that the lower portion is in contact with the peat. It is customary to give cuttings a spacing of from 10 to 12 inches each way.

Citrus.—In the few states where citrus is grown on a commercial basis the general method of propagation follows a rather definite procedure. The difference in soils and other environmental factors makes the choice of stock a consideration of primary importance.

FIG. 143.—Sour orange seedlings in seedbed.

1. *Stocks.*—The use of the proper stock for citrus is an important factor in determining the ultimate success of the orchard. The *sour orange* is by far the most widely used stock at the present time. It is resistant to foot rot and gum disease and will grow in soils where these diseases prevent the growth of other species. The sour orange makes a good union with grapefruit and orange and appears entirely compatible with both; but it does not give good results as a stock for lemons, nor is it compatible with Satsuma orange.

Sweet orange is not so resistant to foot rot as the sour orange, but where disease is not a limiting factor it is a very satisfactory stock. At the present time, it has become the standard stock for lemons in California. It apparently makes a good union with all common commercial forms of citrus, but in most cases the sour orange continues to rank above it. *Rough lemon* is a very vigorous stock, showing its superiority in growth in sandy soils of low fertility. The fruit produced on this stock is somewhat coarse and its use is limited to sections where the other stocks do not make sufficient growth. *Grapefruit* may also be used as a stock, since it is a comparatively vigorous grower and shows a high degree of compatibility. On the other hand there is no especial merit in its use, and it is seldom considered.

The *trifoliate orange* is the standard stock for the entire Gulf coast, outside the areas of commercial production. It is widely used along the coast from Corpus Christi, Tex., to northern Florida. It is universally used as a stock for the Satsuma, which does not succeed on sour orange; trifoliata is commonly used, also, for kumquat and Ponderosa lemon and for any of the grapefruit and round oranges in the coastal region mentioned. Trifoliata is a deciduous species and is hardy to cold. It is frequently stated that it imparts this cold resistance to the top budded on it, but there is no reason to think that this may be true. Many other species of citrus have been tested for their value as stocks, and others are now on trial. *Calamondin* has been tried with indifferent results, and many hybrids, especially *citrange* and *tangelo*, have shown considerable promise.

Identification of the various stocks may be made by means of leaf characters, but in many cases no leaves are present on the stock. Certain chemical tests have been devised to distinguish sweet orange, sour orange, rough lemon, and grapefruit. A knowledge of the general appearance of the seed of the various stocks is of considerable value when a seedbed is to be planted.

Citrus seed is nearly always planted in a special seedbed, where better germination may be obtained and better attention given to the young plants. Most of them will need protection against scab. The young plants at the end of a flush of growth will be 8 to 12 inches high. They are dug with a spading fork; all weak, crooked, or otherwise undesirable plants are culled out; and the good plants set in the nursery row about 15 inches apart, the rows being 3 to 4 feet apart. They are allowed to attain a diameter of $\frac{1}{4}$ to $\frac{3}{8}$ inch before they are budded.

2. *Budding.*—Budding of citrus is similar in most respects to that of peaches. The T bud is used, although many nurserymen

prefer the inverted T. The two main points of difference are in the cutting of the bud from the stick with a piece of wood attached,

FIG. 144.—Sour orange seedlings lined out and ready to bud.

instead of slipping it off, as in peach; and in tying, which is done with a short strip of waxed cloth. Many budders wrap the cloth

FIG. 145.—Budded citrus trees, staked for training.

or tape over the bud itself, while others leave the bud exposed. Citrus may be budded at any time the bark will slip. When expensive buds are being used, the nurseryman feels safer when spring budding is done and the danger of loss from cold is eliminated.

FIG. 146.—Budded citrus trees, ready to remove from nursery.

3. *Training of Tops.*—The buds are forced and the old tops removed, as is the common practice in all nursery work. The

FIG. 147.—Citrus trees, balled and burlapped, ready to plant in the orchard.

training of the young shoot receives careful attention, however, from the time it first starts out. It is tied at close intervals to a 1- by 1-inch stake, and a straight shoot is secured. The common practice is to allow this shoot to reach a height of about 36 inches, then to head it back to 30 inches, and finally to force out a few branches to develop the permanent framework. Citrus nursery trees differ from most others in this respect: they are usually sold as two-year-old trees, with tops already developed, instead of

FIG. 148.—Trifoliate orange, used as stock for Satsuma and other types of citrus in colder sections of the Gulf coast.

being sold as one-year "whips" like peaches or other deciduous trees. This is one reason, aside from the fact that they are evergreens, why citrus trees are usually sold as balled and burlapped (B & B) and seldom bare-rooted.

4. *Own-rooted Plants.*—The production of the various species of citrus on their own roots is desirable for certain purposes. Uniform stocks might be produced, not only for experimental use but for commercial orchards as well. Some species and varieties, for

which no stock of satisfactory performance has been found, are best propagated as own-rooted plants.

Cuttings of most of the species of citrus, and some related genera, have been rooted in California. Stem cuttings 4 to 6 inches long with about seven nodes have given the best results. The lower leaves were removed, but the three or four upper ones were left. Rooting percentages of 75 per cent and upward were obtained.

The Meyer lemon has been produced on its own roots by setting young plants, budded on sour orange, in a deep furrow. After the plants have grown well above the surface, they are partly girdled at the bud union and the furrow is filled.

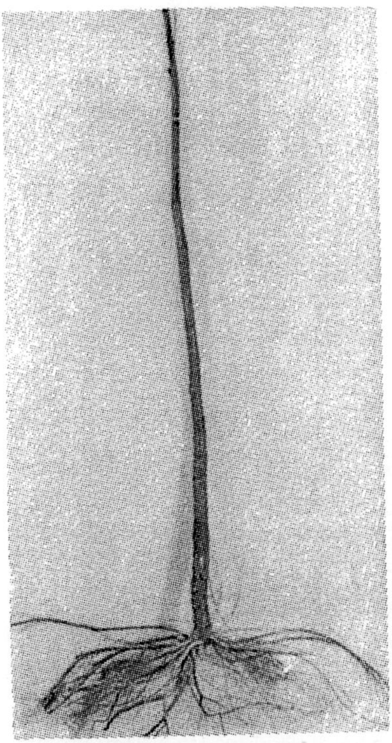

FIG. 149.—Fig tree grown from cutting, showing one year's growth.

Layerage has been used successfully in South Africa for the production of orange trees. In a seventy-year-old grove, layered trees were reported to be as large as comparable seedlings and were reported to be the best and most prolific trees.

Fig.—This fruit has always been found easy to propagate, and nursery trees can be produced on a large scale at very little expense.

1. *Cuttings.*—The common method of propagating figs is by means of stem cuttings, which are usually made from wood of the previous season. Wood with comparatively short internodes is preferable to that in which they are extremely long. The cutting should have a minimum of two nodes, but those of the proper length, approximately 8 inches, will have three or more nodes. The terminal portions of the shoots may be used exclusively for cuttings, but in seasons of moderate injury from cold these tips may be killed, when the rest

of the wood is unharmed. In at least one case under observation all cuttings from tips failed to grow, while good results were obtained from those made from basal portions of the same shoots. A satisfactory practice is to remove entire shoots of one-year wood and cut them into proper lengths.

The cuttings of figs should be made in the latter part of the dormant season, usually in February in the South. They are set directly in the nursery row in beds where the soil has been well prepared and where adequate provision has been made for drainage. If the soil permits, the row may be marked off, and the cuttings pushed down into the loose earth. Another method is to open the bed with a turning plow, set the cuttings against the straight side with enough soil pressed against them to hold them in place. The row is then covered by the plow. In either case the cuttings should be set deep in the ground so that only one bud is exposed. In a dry season one irrigation in early spring will be of great value in promoting growth of the cuttings. Frequently 50 per cent of the cuttings will grow, and, under very favorable conditions, a much higher percentage. Close setting of the cuttings, about 8 inches apart in the row, provides a satisfactory stand. These cuttings will make good trees by fall and frequently produce fruit the same season.

The use of suckers or sprouts is frequently observed for replacing a few trees that have died in an earlier planting. It is not advisable to use these plants on a large scale, because their removal is likely to damage the parent plant.

2. *Budding.*—Top-working of figs may be done with little difficulty. Work in California has shown that young fig trees may be budded easily with the T bud. The same method is successful also on large trees, in wood one to three years old, provided budwood of large diameter is available. This method has been used successfully at various times during the growing season whenever the bark was slipping. Patch buds have also been used with good results. It is advisable in either case to place the buds early in the season, using storage budwood, and force them out, in order to get the maximum amount of growth.

3. *Grafting.*—The cleft graft and bark graft are used more commonly in top-working of figs. Cleft grafting is successful on branches and trunks 3 to 4 inches in diameter. The work is done in late winter or early spring, before growth starts. Bark

grafting is done in March or April after the bark has begun to slip. This method, and also the preceding one, are carried out in accordance with general instructions previously mentioned. It is advisable in either case after the wax has been applied to cover the stubs and scions with whitewash to prevent sunburn. The use of paper sacks as coverings is also recommended to prevent drying of the scions. All grafts, but more specifically bark grafts, require staking or bracing to prevent their being damaged by wind.

4. *Seed.*—In the case of varieties that produce viable seed this method of propagation can be followed. Its use, however, is restricted almost entirely to breeding work. Artificial pollination of varieties that are ordinarily sterile has been accomplished in Texas in recent years as part of a program of breeding.

Avocado.—Rootstocks for the avocado are chosen from three races: Mexican, Guatemalan, and West Indian. In California, seedlings of the Guatemalan type have given better results than those of the Mexican; there are, however, some individual trees of the Mexican race that produce very desirable seedlings. Hybrids of the two races have been very satisfactory. In Florida, seedlings of the West Indian type have been used to a great extent, but hybrids with the Guatemalan are showing promise. For Texas conditions the Mexican seedlings and those of the Guatemalan do not give such good results as do the West Indian seedlings. The Mexican type is most resistant to cold, the Guatemalan intermediate, and the West Indian most tender. Resistance to wind and to soil alkali are other factors that must be considered.

1. *Budding.*—The general method of propagating the avocado has been by means of the T or shield bud. This method has not been very satisfactory; good budwood is difficult to obtain, many of the buds abscise, and the percentage of success is not high. It is believed that better results can be secured with the patch bud, but this would not overcome the difficulty of poor budwood and blind buds.

2. *Grafting.*—Grafting has become more popular in recent years, especially in Florida. Good graftwood may be secured more readily than good budwood, and the difficulty with blind buds is lessened. The side graft has been used with great success on rather small seedlings. Whip grafts have also been used, but the removal of the top lessens the chances for success. Inarching of seedling stocks grown in cypress boxes is a rather common

practice. A modified cleft graft, in which a sawcut is made before the stock is split, has given good results on larger stock.

3. *Top-working.*—The modified cleft graft is used also in top-working larger trees. The stub and the cleft are covered with grafting wax, and a strip of heavy paper is wrapped and tied around the stub. The paper, which extends above the top of the

FIG. 150.—Avocado grafted by approach—now on seedling rootstock.

scion, is filled with sand and peat moss, which is kept damp. Many avocado trees have been top-worked successfully by this method within recent years.

4. *Cuttings.*—Recent experiments with growth-promoting substances have not produced satisfactory rooting of commercial varieties of avocado, although stem cuttings of two-year Mexican nursery seedlings were rooted. Stem cuttings of Fuerte grafted on Mexicola (Mexican) root united with the root and produced

roots on the stem portion. Leafy-twig cuttings of Fuerte collected in December were rooted in a glass chamber covered with cheesecloth.

Rose.—The rose has been a popular flower since the earliest periods of history. Its great range of forms and colors adapt it to a wide variety of uses. Methods by which roses are prop-

FIG. 151.—An avocado tree top-worked by the modified cleft graft.

agated vary with the species and the geographic locality; seeds, cuttings, layers, suckers, buds, and grafts are used.

1. *Seeds.*—New varieties of roses originate largely as selections from seedlings. Occasionally, one originates as a sport. Some true species, as *Rosa multiflora*, may be economically reproduced from seeds. The rose fruits, or "hips," should be collected as soon as ripe and the seed removed. From this time until planted the following spring it is important that the seed be kept con-

tinuously moist. Immediate stratification of the seed in sand or some other moist material is recommended. It is important that they never be allowed to become dry. Experiments indicate that the best temperature during stratification is about 41°F. The seeds of some species apparently have a longer rest period than

FIG. 152.—Cutting multiflora rose propagation material into uniform lengths of 8 inches. The cuttings are placed in the box on the bench for convenience in tying into bundles of 200 each. (*Courtesy A. F. Watkins, Dixie Rose Nursery, Tyler, Tex.*)

others. Stratification at 41°F. for 270 days is recommended for those of the dog brier (*Rosa canina*). Ninety days is sufficient for the Pasture rose (*R. humilis*) and 50 for the multiflora (*R. multiflora*). Stratification serves to keep seed from drying out and thus preserves their viability, and cold temperature breaks the rest period. Seed stratified in late summer or early fall are commonly planted the following spring in either nursery rows or special seedbeds. Near the end of the first growing season they are ready to be budded. If they are not to be budded or grafted the seedlings may be transplanted to their location at the end of the first growing season.

2. *Cuttings.*—In general, roses grow readily from cuttings, and no other method is followed so extensively. Cuttings are used in the multiplication of named varieties and also for the starting of stocks for budding or grafting.

Rose wood in various stages of maturity may be used for cuttings. A comparatively simple method of increasing plants

FIG. 153.—The flat iron on the front of this implement smooths the row; the revolving disc blade opens the furrow; and the roller in the rear with rods 8 inches apart marks the places for planting rose cuttings. (*Courtesy A. F. Watkins, Dixie Rose Nursery, Tyler, Tex.*)

of some species is by means of *dormant hardwood cuttings* made in late fall or winter. They may also be made very early in the spring, though the practice is not encouraged. The cuttings, usually made from canes about the size of a lead pencil, are cut into lengths from 6 to 8 inches. It is important that the basal and terminal cuts be made immediately below and above nodes, respectively. Removing all buds except the one at the top node, a practice known as *disbudding*, tends to discourage the development of water sprouts later in the growth of the plant. It is important that the bottom of the cutting is placed in the soil and that it is not planted in an inverted position. Uncertainty as to which is the bottom end may be dispelled by examination of the leaf scar, which may be seen ordinarily on the stem beneath each bud.

Two general practices are followed in the planting out of hardwood cuttings. (*a*) Where the winters are severe they are commonly planted in a cold frame or other similar structure. Sand, sandy loam, or, in some cases, peat, either with or without sand, is used for the rooting medium. By the time growth starts in the spring the cuttings will have formed roots and

FIG. 154.—Rose cuttings are planted in the rows by hand, after which the soil is pressed closely about the cuttings. (*Courtesy A. F. Watkins, Dixie Rose Nursery, Tyler, Tex.*)

may then be transplanted either to nursery rows in the field or to the place where they are to grow permanently. (*b*) By another practice, which is somewhat prevalent in sections where the winters are mild, hardwood cuttings after having been disbudded are planted directly into the nursery rows in the field. This is usually done during the latter part of January or February. The rows are usually 6 feet apart, and cuttings are spaced about 10 inches apart in the row. Better rooting will be obtained if the cuttings are planted in a well-drained sandy or sandy-loam soil. Drainage may be facilitated by planting on beds from 4 to 6 inches high. By the time top growth starts in the spring the cuttings will have formed roots. The plants that result develop to a size suitable for budding by June of the same year. At that time they may be budded, but more commonly budding is delayed until August. These phases of propagation will be discussed in a paragraph that follows. Cuttings that are not to be budded, for example, those named varieties that are to produce "own-rooted-plants," are allowed to grow in the nursery rows one or two seasons.

Fruit jars are sometimes inverted over hardwood rose cuttings that are planted about the house or yard. The beneficial effect of these is probably due to the maintenance of high humidity about the cutting. Commercial growers accomplish the same results by growing the cuttings under a glass sash or cover and by watering them at frequent intervals.

In some sections the rooting of hardwood cuttings is not entirely successful. Another practice followed to some extent is to grow the rose plants from *softwood cuttings* made from material during its current or first season of growth. Such material is soft, immature, and succulent, and cuttings made from it require more care and attention in the propagation bed. The cuttings, 6 to 7 inches long with only the basal leaves removed, are planted upright about 3 inches deep. They are usually set in a special propagation bed with sand as the rooting medium and are shaded with cheesecloth until rooted. Slatted frames offer the same kind of protection. Frequent waterings are necessary to keep the leaves from wilting and dropping. These cuttings root commonly in from 10 to 14 days—a shorter period than required for hardwood cuttings. As new growth starts, indicating root formation, the shade is gradually removed, first for only a part of each day and finally for the entire day. Softwood cuttings should be made as early in a season as the wood becomes sufficiently mature. In the South, April or early May is not too early; in the North they are made the latter part of July and August. If made early they form a better root system and have a longer period during which to mature. Cuttings may be taken from any part of a current-season shoot. Those taken from the tips of growing canes root more quickly than others taken from the base or middle portion. The difference, however, is very slight and, since plants started from the older parts of canes mature more properly in the fall, many propagators prefer to use the older material. One serious objection to softwood cuttings is that they cannot be disbudded easily. At the end of the first season, plants from softwood cuttings may be used as lining-out stock; in the case of named varieties the plants may be moved to their permanent location.

3. *Stocks.*—Most species of roses are congenial when intergrafted or interbudded. Theoretically, then, almost any species could be used as a stock. Different species, however, vary widely in the many characteristics that determine their merits for stock purposes. Some species, for example, root readily from cuttings; others, very poorly. Others, although they grow readily from cuttings, form weak and poorly developed root systems. Likewise there are differences in such characteristics as suckering, tolerance of or resistance to nematodes, and the presence or absence of thorns. In actual practice, only a few species are used for stock purposes. Thornless strains of *Rosa multiflora* are the most popular stocks. Cuttings of it can be handled more easily, and the resulting plants can be budded more readily than is the case with species that have thorns. The species roots readily, forms a good root system, and is decidedly resistant to nematodes. It seldom forms suckers. The dog brier (*R. canina*) is an excellent rose stock but has very heavy thorns that make it difficult to handle. It has a tendency to form suckers in some regions. *Rosa manetti* and *R. odorata* are the principal stocks used for greenhouse roses. *Rosa rugosa* is used as a stock for "tree roses."

4. *Grafting.*—Grafting was long a standard practice in the propagation of greenhouse roses. One-year-old rootstocks grown from cuttings may be whip-grafted with the beginning of the second growing season. The stocks may be dug and the work done indoors during the winter, or they may be grafted in the field in early spring. In either case plants of commercial planting size are produced in one season.

Grafting is one method of propagating rose plants for forcing purposes. The plants are grafted in midwinter and carried along in small pots until sold by the propagator in the late winter or early spring. The plants may be held for a time in pots by the grower, but they are usually planted in the greenhouse benches by midsummer for production of cut flowers the following winter. In sections where winters are mild, rose plants may not become completely dormant, in which case it is difficult to obtain scions suitable for grafting. The susceptibility of grafted plants to the attack of crown gall discourages the practice of grafting roses.

5. *Budding.*—The T, or shield, bud is the method followed quite universally in budding rootstocks grown from seed or cuttings. Buds are taken from bud sticks of the current season's growth. It is important that the scions be cut early in the morning, that the leaves be removed by clipping leaf petioles ⅓ inch out from bud, and that the bud sticks be kept in a cool, moist place until used. A fresh supply of budwood should be

FIG. 155.—This field of roses was budded in August. The tops are being cut the following spring to force the dormant buds into growth. (*Courtesy A. F. Watkins, Dixie Rose Nursery, Tyler, Tex.*)

cut each day whenever it is practicable to do so. If insulated in moist moss or other material and kept at a temperature of 32 to 40°F. they may be safely used over a period of 4 or 5 days or longer.

Hardwood cuttings planted in the field in late winter or one-year plants lined out in early spring make sufficient growth in warmer parts of the United States to be budded in early summer. Budding that is done in June is referred to as *June budding*, and plants that are produced thereby are known as *June buds.* June-budded rose plants may make from 18 to 24 inches of growth before the end of the first season. They are not likely, however, to make sufficient growth to be classed as No. 1 plants. An advantage of June budding is that the grower is able to produce a rosebush suitable for planting within one growing season.

Instead of budding the rose stock in June the operation may be delayed until late summer. Buds are inserted at or slightly below the ground line. To do this it is necessary to remove a surface layer of soil from around the base of the plant. The buds

FIG. 156.—Roses dormant-budded in August, showing one month's growth the following spring. (*Courtesy A. F. Watkins, Dixie Rose Nursery, Tyler, Tex.*)

are placed in the smooth part of the cutting that was disbudded at planting time and not in the new growth from the cutting. Raffia and rubber bands are commonly used for tying. Twine also may be used. No wax or similar material for excluding air is necessary. The buds unite with the stock but are not forced into growth until the following spring. Since they remain dormant over winter the method is known as *dormant budding.* In winter or very early spring the native top is cut off, and the inserted bud begins growth in early spring as a result of the stimulus. When it has made 6 or 8 inches' growth, it is pruned to encourage branching. Under favorable growing conditions it develops into a large, vigorous plant by the end of the growing

FIG. 157.—Dormant rose buds forced in early spring, showing growth up to June 1. (*Courtesy A. F. Watkins, Dixie Rose Nursery, Tyler, Tex.*)

season. The stock of such a plant has two seasons in which to make its growth, while the top has only one. If as a result of

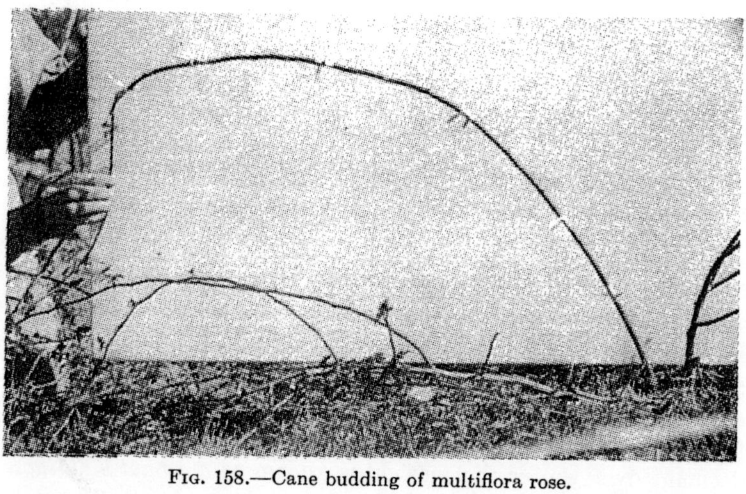

FIG. 158.—Cane budding of multiflora rose.

drought or other unfavorable conditions the rose stocks cannot be budded in late summer, the budding may be done early the following spring. The buds are forced immediately and produce strong rose plants by the end of the growing season.

A common procedure in propagating roses is to plant rose cuttings and bud them after they have rooted and made some growth. Instead, buds may be inserted in branches of a plant that is to be used later for cuttings. This is known as *cane*

FIG. 159.—*Rosa rugosa* used as a rootstock for tree roses.

budding. As many as 8 or 10 buds spaced 5 to 7 inches apart are inserted in a long cane. The time at which this is done depends on the most favorable season for making the cuttings. The several buds on a cane unite with it, within a period of 2 weeks. Immediately after union, if they are handled as softwood cuttings, and the following winter in case of hardwood cuttings, the canes are cut into as many cuttings as there are living buds. All native buds are removed. When the cutting has rooted and the shoot begins to develop from the bud, it is best that the new shoot be cut back to about three strong buds so that it will make a bushy plant.

Certain species of roses make a very rank growth and produce strong, relatively inflexible canes. With proper pruning, such species can be caused to grow into a single upright stem for 4 or more feet from the ground. When such stocks are budded at a height of 4 or 5 feet above the ground, and the bud develops into a top or bush, the resulting plant is known as a *standard* or *tree* rose. Tree roses are interesting novelties wherever found, and in some sections have value for use in landscaping. Any upright-growing species that has been trained to a single standard can be used as a stock; *Rosa rugosa* is used more commonly than any other. The beauty of the roses will be appreciated more if a variety with a long blooming season is used for a top.

6. *Layers and Suckers.*—Most varieties of roses can be reproduced from layers. The method may be used for the production of a limited number of plants, but it is seldom used in commercial propagation. Likewise, some varieties produce suckers freely and these may be separated from the parent plant and used as a source of new plants.

7. *Own-rooted, and Budded and Grafted Roses.*—Rose plants that are allowed to grow to maturity from cuttings are known as own-rooted plants. Those that might be grown to mature size from seed likewise would be regarded as own-rooted. In general, roses grow readily from cuttings of one kind or another. It would seem, then, that there would be no occasion for growing any kind other than own-rooted plants. Some few varieties, however, do not root well. Others, although they grow from cuttings, form weak root systems that are not adapted to certain soils, or perhaps are not tolerant of insects that are prevalent in some soils. Certain varieties on their own roots are perfectly acceptable for some conditions, while under a different set of conditions they respond very poorly.

Budded or grafted rose plants consist of two components: the stock and the top that develops from the bud or graft inserted in the stock. The propagator can select a stock and scion each with certain desirable characteristics and by budding or grafting grow a plant that is more useful in many cases than would be possible with own-rooted plants.

Ligustrum.—The ligustrums are valuable ornamental plants; the different species are grown in most sections of the United States, and interest in the propagation of them is widespread. The *California* and *Amur* privets may be propagated readily from either softwood or hardwood cuttings and also from seed. Cuttings may be started in a specially prepared bed with sand

FIG. 160.—A Japanese ligustrum grafted on an Amur ligustrum rootstock.

or a mixture of sand and peat for the rooting medium, and the plants later moved to the nursery. Hardwood cuttings are frequently planted directly in the field, with variable results. It was very early shown that treating California privet cuttings with potassium permanganate stimulated rooting. *Japanese* (*Ligustrum japonicum*) and *Waxleaf* (*Ligustrum lucidum*) ligustrum cuttings do not root so readily, and these species are frequently budded or grafted on either California or Amur stock. Chip or T buds are set, usually on California stock, and forced either the same or the following season, depending on the time of year when the budding is done. Scions may also be whip-grafted into grafting stocks. This is done in late winter if the work is done indoors, or in early spring if the stocks are

FIG. 161.—Lilac grafted on California privet rootstock.

grafted in the nursery. Bench grafts are replanted in the nursery after they callus. Variegated forms of ligustrums occur frequently, and these may likewise be readily propagated by budding and grafting. Incidentally, California privet is frequently used as a budding or grafting stock for horticultural varieties of lilac. The steps by which the plants are produced are the same as outlined for Japanese and Waxleaf ligustrums.

In the replanting of the lilac plants, however, it is important that the union be placed below the ground level so as to encourage rooting from the lower portion of the lilac scion.

Conifers.—The widespread use of the coniferous evergreens in ornamental plantings has attracted considerable attention to the propagation of such species as arborvitae, retinospora, juniper, and yew, as well as pine, spruce, and cypress. A number of these plants may be propagated by cuttings, and in a few cases grafting has been used successfully.

1. *Cuttings.*—The best time for making cuttings of most of these species is in late fall or early winter. Short tips of 3 to 4 inches are used, although some prefer longer cuttings. In some cases they are made with "heels," but this is not essential. The lower half of the cutting, with foliage removed, is placed in sand in the greenhouse bench, and the bed watered to firm the sand. Shading is usually practiced to some extent, and a rather high humidity maintained. Root formation usually occurs in spring, but some of the junipers require a year or even longer to form roots.

Summer cuttings of immature wood have also been used with some success in the case of arborvitae and similar species. As in the case of the deciduous softwood cuttings, the conditions of temperature and humidity must be maintained at the optimum to secure best results. Frames of glass are used to maintain a high degree of humidity, and sand has been found to be a satisfactory medium.

In both methods described, the use of peat and sand has given good results, and the use of peat alone has given especially good results with some of the more difficult species, such as the junipers.

2. *Grafting.*—Although grafting of the conifers is recognized as a rather difficult process, it has been used successfully in the propagation of named varieties of spruce and pine. The seedlings are potted in fall and brought into the greenhouse, where they are grown until grafting is done in January or February. The veneer graft is more commonly used, but a modified side graft has also been successful. In either case the seedling top is left intact. The pots are set inside a frame, with sphagnum moss packed around them. The moss is kept damp and the case closed to maintain humidity. Part of the seedling top is

removed in 3 to 4 weeks, and in the same length of time again the rest of the top may usually be removed.

3. *Nursery Methods.*—Coniferous evergreens grown from either cuttings or grafts are usually carried through one season in pots in frames or lath houses. The plants are then lined out in the nursery row and grown until they reach suitable size for transplanting. A common practice is to remove alternate plants rather early, as a thinning process, to give the others a better opportunity to develop. Another thinning may be made later to produce larger specimens. All the evergreens are transplanted as balled-and-burlapped stock.

Questions

1. What is the objection to grafting peaches?

2. Name the different rootstocks used for peaches. What are the merits of each?

3. How is the Marianna plum rootstock propagated?

4. How are peach seed treated in order to improve germination?

5. What method of budding is used most in propagating the peach? What kind of budwood is used?

6. Outline the successive steps, with dates of each, in propagating a June-bud peach tree; a one-year-old peach tree.

7. What methods may be used in top-working a large peach tree?

8. What rootstocks are used for plums? Apricots? Cherries? Almonds?

9. What are the causes of difficulties in getting seeds of cherry to germinate?

10. How should apple seed be handled in order to insure good germination?

11. Outline the various methods used in propagating apple trees suitable for orchard planting.

12. Tell how to top-work a large apple tree.

13. How are dwarf apple trees produced? Of what value are they?

14. What rootstocks are used for pears?

15. Name the various methods of budding and grafting used in propagating the pear.

16. Tell how to produce a pear tree with blight-resistant framework; a dwarf pear tree.

17. How should pecan seed be treated in order to insure good germination?

18. What method would be followed in budding a small pecan tree? In budding one 6 inches in diameter? In top-working one 15 inches in diameter?

19. What methods are used in propagating persimmons?

20. List the different ways of propagating grapes.

21. Grape plants that are grown from cuttings are used for what purposes?

22. How are grape rootstocks produced?

23. What are the requirements of a good rootstock for grapes?

24. Outline the different ways by which grape plants suitable for planting in the vineyard are propagated.

25. How are blackberries propagated? Dewberries? Raspberries? Strawberries? Blueberries?

26. What different rootstocks are used for the various kinds of citrus?

27. Outline the steps in producing nursery trees of citrus.

28. What different methods are used in propagating figs?

29. Outline the different ways of handling hardwood cuttings of rose.

30. What are the requisites of a good rose stock?

31. What is cane budding? How are "tree roses" propagated?

32. What methods are used in propagating California privet? Cherry laurel? Japanese Flowering quince? Lilac? Waxleaf ligustrum? Conifers?

Suggested References

Peach—Stone Fruits

1. HANSON, C. J., and E. R. EGGERS: Propagation of Fruit Plants, *Calif. Agr. Ext. Serv. Circ.* 96, 1936.
2. HEPPNER, MYERS J.: Grafting Affinities with Special Reference to Plums, *Calif. Agr. Exp. Sta. Bull.* 438, 1927.
3. HOWE, G. H.: Mazzard and Mahaleb Root Stocks for Cherries, *N. Y. (Geneva) Agr. Exp. Sta. Bull.* 544, 1927.
4. HUTCHINS, LEE M.: Nematode-resistant Peach Root Stock of Superior Vigor, *Proc. Amer. Soc. Hort. Sci.*, **34**: 330–338, 1937.
5. TUFTS, W. P., and L. H. DAY: Nematode Resistance of Certain Deciduous Fruit Tree Seedlings, *Proc. Amer. Soc. Hort. Sci.*, **31**: 75–82, 1934.
6. TUKEY, H. B.: Plant Propagation, *N. Y. (Geneva) Agr. Exp. Sta. Circ.* 138, 1934.
7. VAN ALSTYNE, L. M.: The Plum in New York, *N. Y. (Geneva) Agr. Exp. Sta. Circ.* 134, 1932.
8. YERKES, GUY E.: Propagation of Trees and Shrubs, *U. S. Dept. Agr., Farmers' Bull.* 1567, 1929.

Apple—Pome Fruits

1. ALLEN, F. W.: Apple Growing in California, *Calif. Agr. Exp. Sta. Bull.* 425, 1927.
2. CHILDS, LEROY, and GORDON G. BROWN: A Study of Tree Stocks in Relation to Winter Injury and Its Prevention, *Ore. Agr. Exp. Sta. Circ.* 103, 1931.
3. GOULD, H. P.: Apple Growing East of the Mississippi River, *U. S. Dept. Agr., Farmers' Bull.* 1360, 1924.
4. REIMER, F. C.: Blight Resistance in Pears and Characteristics of Pear Species and Stocks, *Ore. Agr. Exp. Sta. Bull.* 214, 1925.
5. ROBERTS, R. H.: Factors Affecting the Variable Growth of Apple Grafts in the Nursery Row, *Wis. Agr. Exp. Sta. Research Bull.* 77, 1927.

6. ———: Some Stock and Scion Observations on Apple Trees, *Wis. Agr. Exp. Sta. Research Bull.* 94, 1929.
7. Swarbrick, Thomas, and R. H. Roberts: The Relation of Scion Variety to Character of Root Growth in Apple Trees, *Wis. Agr. Exp. Sta. Research Bull.* 78, 1927.
8. Yerkes, Guy E.: Propagation of Trees and Shrubs, *U. S. Dept. Agr., Farmers' Bull.* 1567, 1929.

Pecan and Walnut

1. Bailey, J. E., and J. G. Woodroof: Propagation of Pecans, *Ga. Agr. Exp. Sta. Bull.* 172, 1932.
2. Blackmon, G. H.: Topworking Pecan Trees, *Fla. Agr. Exp. Sta. Bull.* 170, 1924.
3. Brison, F. R.: The Storage and Seasoning of Pecan Bud Wood, *Tex. Agr. Exp. Sta. Bull.* 478, 1933.
4. Reed, C. A.: Nut-Tree Propagation, *U. S. Dept. Agr., Farmers' Bull.* 1501, 1926.
5. Sitton, B. G.: Vegetative Propagation of the Black Walnut, *Mich. Agr. Exp. Sta. Tech. Bull.* 119, 1931.

Grape

1. Husmann, George C.: Testing Phylloxera-resistant Grape Stocks in the Vinifera Regions of the United States, *U. S. Dept. Agr., Tech. Bull.* 146, 1930.
2. Jacob, H. E.: Propagation of Grapevines, *Calif. Agr. Ext. Serv. Circ.* 101, 1936.
3. Winkler, A. J.: Some Factors Influencing the Rooting of Vine Cuttings, *Hilgardia*, **2**: 329–349, 1927.

Berries

1. Colby, A. S., H. W. Anderson, and W. P. Flint: Bramble Fruits, *Ill. Agr. Exp. Sta. Circ.* 427, 1937.
2. Darrow, George M.: Dewberry Growing, *U. S. Dept. Agr., Farmers' Bull.* 1403, 1926.
3. Hartman, Henry: The Cane Fruit Industry in Oregon, *Ore. Agr. Exp. Sta. Circ.* 48, 1923.
4. Hendrickson, A. H.: Small Fruit Culture in California, *Calif. Agr. Exp. Sta. Circ.* 164, 1917.
5. Waldo, George F., and O. T. McWhorter: Crown Division—A Means of Propagating Everbearing Strawberries, *Ore. Agr. Ext. Serv. Bull.* 488, 1936.

Citrus

1. Camp, A. F.: Citrus Propagation, *Fla. Agr. Exp. Sta. Bull.* 227, 1931.
2. Traub, H. P., and W. H. Friend: Citrus Production in the Lower Rio Grande Valley of Texas, *Texas Agr. Exp. Sta. Bull.* 419, 1930.

3. WEBBER, H. J.: Variations in Citrus Seedlings and Their Relation to Rootstock Selection, *Hilgardia*, **7**: 1–79, 1932.
4. ———: Influence of Rootstock Strains on Yield and Size of Lemons, *Proc. Amer. Soc. Hort. Sci.*, **31**: 83–88, 1934.

Fig

1. CONDIT, I. J.: Fig Culture in California, *Calif. Agr. Exp. Sta. Circ.* 77, 1933.
2. GOULD, H. P.: Fig Growing in the South Atlantic and Gulf States, *U. S. Dept. Agr., Farmers' Bull.* 1031, Rev. 1935.
3. STANSEL, R. H., and WYCHE, R. H.: Fig Culture in the Gulf Coast Region of Texas, *Tex. Agr. Exp. Sta. Bull.* 466, 1932.

Avocado

1. HAAS, A. R. C.: Propagation of the Fuerte Avocado by Means of Leafy-Twig Cuttings, pp. 126–130, *Calif. Avocado Assn. Yearbook* 1937.
2. ———: Further Progress in the Rooting of Fuerte Avocado Cuttings, pp. 130–132, *Calif. Avocado Assn. Yearbook* 1937.
3. HODGSON, ROBERT W.: The California Avocado Industry, *Calif. Agr. Ext. Serv. Circ.* 43, 1934.
4. WOLFE, H. S., L. R. TOY, and ARTHUR L. STAHL: Avocado Production in Florida, *Fla. Agr. Exp. Sta. Bull.* 272, 1934.

Rose

1. TUKEY, H. B.: The Propagation of Multiflora Rootstocks for Roses by Soft Wood Cuttings, *N. Y. (Geneva) Agr. Exp. Sta. Bull.* 508, 1931.
2. WEINARD, F. F., and H. B. DORNER: *Rosa Odorata* as a Grafting Stock for Indoor Roses, *Ill. Agr. Exp. Sta. Bull.* 290, 1927.
3. ——— and S. W. DECKER: Summer-budded versus Winter-grafted Roses, *Ill. Agr. Exp. Sta. Bull.* 358, 1930.
4. ——— and ———: Effects of Prolonged Storage on Forcing Qualities of Summer-budded Roses, *Ill. Agr. Exp. Sta. Bull.* 409, 1934.

Ornamentals

1. CHADWICK, L. C.: Studies in Plant Propagation, *N. Y. (Cornell) Agr. Exp. Sta. Bull.* 571, 1933.
2. HITCHCOCK, A. E., and P. W. ZIMMERMAN: Relation of Rooting Response to Age of Tissue at the Base of Greenwood Cuttings, *Contrib. Bouce Thompson Inst.*, **4**: 85–98, 1932.
3. ——— and ———: Effect of Growth Substances on the Rooting Response of Cuttings, *Contrib. Boyce Thompson Inst.*, **8**: 63–79, 1936.
4. LAURIE, ALEX: The Use of Peat in the Greenhouse, *Mich. Agr. Exp. Sta. Sp. Bull.* 194, 1930.
5. WYMAN, DONALD, and MAURICE W. NIXON: Electric Hotbeds for Propagating Woody Cuttings, *N. Y. (Cornell) Agr. Exp. Sta. Bull.* 618, 1934.
6. YERKES, GUY E.: Propagation of Trees and Shrubs, *U. S. Dept. Agr., Farmers' Bull.* 1567, 1929.

CHAPTER XV

THE RELATION OF PROPAGATION PRACTICES TO DISEASES

Abnormal conditions of plants may be caused by (1) parasitic organisms, (2) viruses, and (3) peculiar physiological disturbances, such as an improper combination of stock and scion.

In some cases the agent that causes the disease is actually disseminated in the various steps involved in propagation; in others the disease may be avoided or rendered less serious by the use of the proper stock and scion. Some of the troubles encountered more or less commonly and requiring consideration are discussed in the following paragraphs:

Root Knot.—A so-called "root-knot disease" of many fruit trees, shrubs, and vegetables is caused by the nematode known as *Heterodera marioni*. It is a minute, slender parasitic worm that lives almost exclusively within the roots of certain susceptible plants. The larvae may be present in a soil, from which they enter young rootlets, or the eggs may hatch within infested roots and the resulting larvae work their way through the plant tissue some distance from the place where they were hatched. In either case the presence of the larvae causes abnormal cell growth in the immediate vicinity, and this results in swellings that become parts of the roots, known commonly as *root knots*. The swellings are usually small, seldom being larger than a garden pea. The number of swellings depends largely upon the number of nematodes present, and that, in turn, may vary with the root system, some plants being more susceptible than others.

Nematodes are distributed widely throughout the warmer regions of the United States and are found throughout the tropical regions of the world. In the Southern states, nematodes are active over a much longer period than along the northern range of infestation. They are thought to be more troublesome in sandy soils than in heavier types.

Once established, nematodes are very difficult to eradicate. A new orchard should be planted only on soil known to be free from nematodes; only trees or shrubs free from nematodes should be planted; and, if possible, species with resistant root systems should be used.

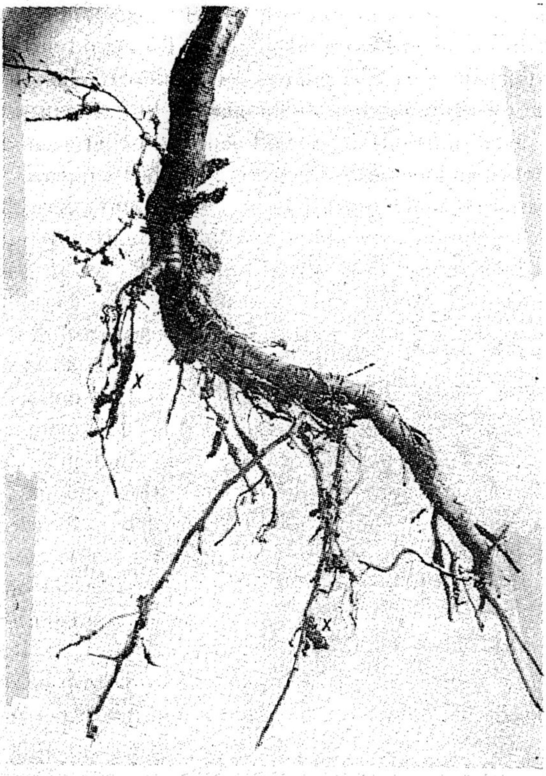

FIG. 162.—Nematode injury on root of peach tree. Characteristic root-knot injury indicated at *X*.

Workers at the University of Tennessee Agricultural Experiment Station grouped plants on the basis of their susceptibility. The following classification is derived from their report and also from the work of other investigators.

Root-knot injury is not to be confused with other swellings that occur on roots of plants. Nodules caused by nitrogen-producing organisms on legumes are attached to roots on one side only and can be detached easily. Root knots, that is,

SUSCEPTIBILITY OF FRUIT PLANTS TO NEMATODE INJURY

Badly infested	Slightly infested	Seldom infested or highly resistant
Fig	Blackberry	Apple
Peach (except on selected stock)	Dewberry	Apricot
Myrobalan plum	Raspberry	Marianna plum
Black walnut	Strawberry	Currant
English walnut	Pear	Gooseberry
All varieties of vinifera grapes	Persimmon	Shalil peach
Almond	Quince	
	Dog Ridge grape (*Vitis champini*)	

swellings caused by nematodes, on the contrary, become part of the root. The grape root louse (phylloxera) and crown gall bacteria both likewise produce characteristic swellings and will be discussed in later paragraphs.

The Grape Root Louse.—The colonists who settled America assumed that standard varieties of European grapes would grow readily in this country because of the luxuriant growth of native grapes. For nearly 200 years they attempted to grow European varieties, and their efforts comprise one of the most interesting, yet tragic, chapters in the history of fruitgrowing. Invariably, European grapes planted in Colonial America died of a mysterious "sickness" known since to have been caused by the grape root louse or grape phylloxera. It is a minute insect that injures the grape principally by feeding on the roots, causing lateral enlargements or swellings, which are also known as *nodes* and *tubercles.* These vary in size on different plants and are attached to the root on one side only. This characteristic distinguishes them from the enlargements caused by nematodes and from the much larger galls caused by crown gall. The decay of the swellings caused by grape root louse causes the root system to become diseased and weakened, and this is more injurious than the parasite. The insect is native to parts of the United States east of the Rocky Mountains. In about 1860 it made its appearance in France and soon had become so widely disseminated that it threatened the existence of the grape-growing industry of the European continent. Some measure of control is effected by planting grapes in sandy soils, where vines are often sufficiently resistant to be practically immune. In the main, however, control is strictly a problem for the propagators. European grapes (*Vitis vinifera*) are highly susceptible to phylloxera. Some strains of American species are resistant to it.

In European countries, rootstocks of American species are used generally for the wine grapes. European varieties when grown in America should be grafted onto a native stock known to be resistant. Many varieties of American grapes do not have sufficient resistance to the grape root louse to justify growing them as own-rooted plants, and congenial rootstocks of resistant species should be used for these. Incidentally, resistance of vines to phylloxera depends on inherent resistance, adaptation to soil and climate, and the degree of congeniality existing between them and the varieties to which they are to be grafted. The use of resistant species has introduced other problems such as stock and scion relationships, but the mastery of the insect has been made complete by the use of species of known resistance.

Woolly Aphis.—The woolly aphis is a pest of the apple and pear and is especially serious because it infests the root system. Roots that are infested present a knobby or knotty appearance; and oftentimes the malformations are confused with those caused by other troubles. Woolly aphis are especially troublesome on young trees and frequently are introduced into new areas on planting stock. Nursery trees that show signs of infestation should be soaked in a solution of 40 per cent nicotine sulphate prepared by adding 1 pint to 100 gallons of water.

Some pear rootstocks are damaged more seriously than others by the woolly aphis. Roots of French pears are very susceptible to its attack; some of the Chinese species, including *Pyrus calleryana*, *P. betulaefolia*, and *P. ussuriensis*, are quite resistant; and the Japanese pear has some degree of resistance.

Crown Gall and Hairy Root.—Frequently abnormal cankerous overgrowths develop on roots or crowns of plants. Isolated enlargements form galls that may range in size up to 1 inch or more in diameter; these are usually attached on one side only, though they may surround the root or stem on which they occur. If the affected portion is more or less continuous, it presents an irregularly enlarged corky appearance. Both types of abnormal growth are caused by a bacterial organism, *Pseudomonas tumefaciens*. The organism may be present in the soil, from which it gains entrance into parts of the plant that later show its presence. It may also gain entrance through wounds made during a grafting or budding operation, and its presence and activity stimulate the abnormal type of callusing and growth that forms galls. Galls may be soft and spongy, in which case they generally decay at the end of the growing season; or the

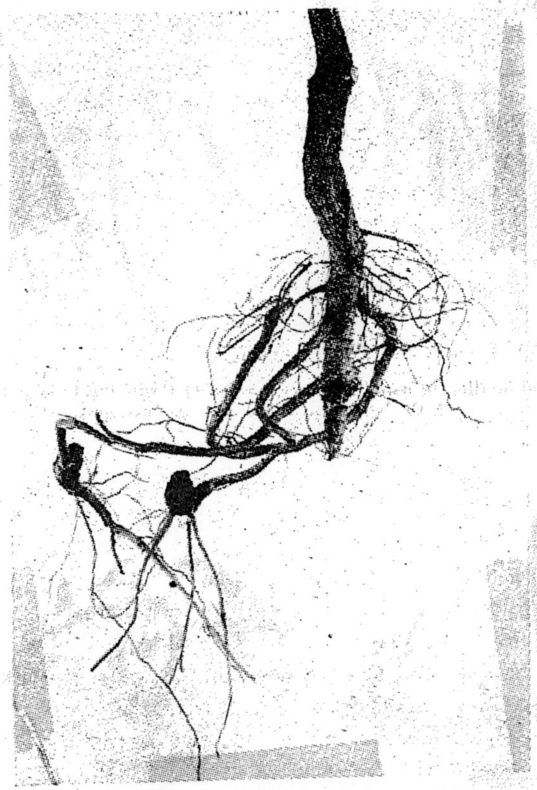

FIG. 163.—Typical crown-gall injury on peach roots.

overgrowth may persist and develop into a woody structure and continue its growth for years. It is doubtful if a plant once infested ever becomes free from the disease. Infested plants usually die after one or two growing seasons, though they may survive in a weakened condition for many years. Species of Rosaceae, including the apple, pear, peach, plum, cherry, almond, rose, blackberry, and raspberry, and many ornamental plants are known to be susceptible. The pecan, walnut, and grape are also known to be affected.

This brief review of the disease and its damage is sufficient to indicate that crown gall is serious. In the handling of nursery stock any plants showing evidences of suspicious overgrowths should be discarded. The practice of using plants that are

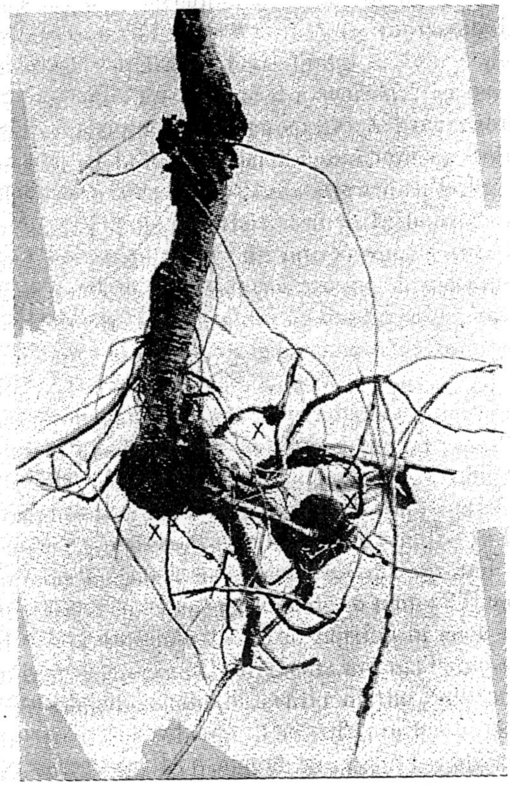

FIG. 164.—Crown gall on main root and branch roots of peach, injury at *X*.

apparently free from disease, but that come from rows or areas in which diseased trees are found, should be discouraged. It is not a safe practice to use plants from which the galls have been removed, even though the wounds are sterilized. It is known that the disease may be disseminated readily by grafting, if diseased material is used. The organism may thus be transmitted from occasional stocks or scions that are diseased to healthy plants. As a precaution to reduce infection, all stocks and scions should be dipped in a disinfectant before grafting. It is well, also, to disinfect the grafting knife at frequent intervals during the course of a day's work. Finally, the grafts may be disinfected again after they are joined together and tied. Mercuric chloride solution, 1 gram to 1,000 cubic centimeters water and 8–8–50 Bordeaux are effective disinfectants for the crown-gall organism.

Hairy root is a disease that is recognized by greatly increased root production from certain areas, which may or may not be swollen or enlarged. The roots are usually small and fibrous in structure. This trouble is closely associated with crown gall and was formerly thought to be produced by the same organism; it is now considered to be caused by *pseudomonas rhizogenes*. The same precautions should be observed in handling nursery stock to prevent the spread of hairy root as for crown gall.

Collar Rot.—A diseased condition that is localized on the trunk of the tree near the surface of the ground is known as *collar rot*. Such a trouble may occur on trees of different kinds and be caused by different agents, such as extreme cold, fungi, and sunscald. If the diseased area completely encircles the tree it becomes girdled and death follows unless the area is repaired by bridge grafting or inarching. Varieties that are especially susceptible to collar rot are frequently budded or grafted onto resistant stocks several inches above the ground line in order to avoid damage. In some cases trees are double-worked in order to provide a resistant splice at the point where damage occurs.

Fire Blight.—This disease is caused by a pathogen, *Bacillus amylovorus*, that may attack the leaves, blossoms, twigs, fruit, and trunk of certain plants. It is known also as *pear blight* because of the very definite susceptibility of the European varieties of pears to the disease. It attacks also the apple and quince and sometimes occurs on the plum, peach, cherry, and rose.

It was known very early that plants of certain species of pears are highly immune to the disease. Among the first attempts to control the disease was grafting of susceptible varieties onto immune rootstocks, with the idea that the resistant stock would render the top likewise immune. It is now known that the rootstock does not influence directly the resistance or susceptibility of a top to the disease. Susceptible varieties of pears growing on resistant rootstocks are no more resistant to the blight than if growing on their own roots. Nevertheless, certain propagation practices may be used to influence the disease indirectly and to render it less serious. One such practice involves the use of a dwarfing stock such as quince. The disease is favored by a succulent and rapid type of growth. Pear trees on quince stock are dwarfed and are not so susceptible to injury by the organism.

Damage from fire blight may be prevented to some extent by top-working susceptible varieties onto resistant stocks in such a manner as to provide a trunk and framework of the resistant species and a relatively restricted bearing surface of the susceptible top. The trunk and main limb of the resistant stock discourage cankers from forming on the framework of the tree and weakening it. The Old Home, Varolosa, and Surprise are blight-resistant varieties that are promising for providing such a framework.

Black End of Pears.—In recent years pear growers on the Pacific coast have been troubled with a peculiar disease known as *black end.* It affects the fruit, causing the calyx end and parts of the flesh adjacent to it to become black. The trouble may appear at any stage of the development of the fruit and cause it to be unmarketable. It has been shown to be associated with certain oriental rootstocks, particularly *Pyrus serotina,* which has been used extensively in an attempt to limit the damage of pear blight. The disease does not spread from tree to tree and is not transmitted by budding or grafting. It has been suggested that it is caused by failure of the rootstock to absorb sufficient water and nutrients from the soil to support normal development of the fruit. Whatever the cause may be, it is clearly a trouble associated with some phase of uncongeniality between stock and scion.

Peach Yellows.—Premature ripening of abnormal fruit, "willowy," slender, much-branched shoots, and pale yellow leaves of reduced size are symptoms of a serious disease known as *peach yellows.* It is a virous disease and has been reported in widely separated peach-growing districts east of the Rocky Mountains but not west of them. Trees may die very soon after first signs of infection, or they may survive for many years. In either case the disease is serious and causes fruit produced to be worthless. Early removal is advocated in order to lessen chances of surrounding trees becoming diseased as a result of infection by leaf hoppers.

It has been established definitely that peach yellows can be transmitted to healthy trees by budding and grafting. Even the smallest bud or graft capable of making union with a stock is sufficient to transmit the disease. It is important that scions to be used in propagation be taken from trees known to be free from this disease. Apparently it is not transmitted in pruning, and it is not probable, then, that it would be communicated from diseased to healthy tissue on a grafting knife or tool.

The symptoms of *little peach* are very much like those of yellows, but the two are considered to be distinct. Little peach, like yellows, can be transmitted to healthy trees by budding and grafting with diseased scions; hence, only scions from trees known to be healthy should be used.

Peach Rosette.—This virous disease of the peach is known in several of the regions of the United States where peaches are grown. Dense tufts of small yellowish-green leaves grow on very short portions of a stem, forming "rosettes." Destroying affected trees is the only control known. Diseased buds and grafts produce diseased trees and of course should be avoided in propagation.

Phony Peach.—This disease is prevalent in Georgia and several adjacent states in which peaches are grown. The area in which it occurs extends into east Texas. Scions affected with the phony disease when grafted or budded on healthy roots produce disease-free trees. But a diseased rootstock results in a diseased tree, even when healthy scions are budded or grafted upon it. Phony peach is especially pernicious because of the delayed inception of the disease. Symptoms do not appear until about 18 months after inoculation.

Peach Mosaic.—In 1931, a serious peach disease was discovered independently by workers in several places west of the Mississippi River. Its occurrence in widely separated districts suggested that it might have been disseminated with nursery trees. Later investigations have shown, conclusively, that buds taken from mosaic-diseased trees will produce trees that are likewise diseased. Symptoms of the disease are small, crinkled, mottled leaves with light-yellow and dark-green mosaic pattern. The internodes are abnormally short; and twigs give rise to a large number of shoots. Fruit is ridged or bumpy, streaked with color, and ripens later than usual. Buds or scions from infected trees will produce diseased trees when budded or grafted onto healthy stocks. A diseased root when grafted on stem or root of a plant will cause the entire plant to become diseased. Peach mosaic will also spread in an orchard from diseased to healthy mature trees, and this indicates that it is disseminated in other ways than in propagation. It is thought to be a virous disease.

Sweet-potato Insects and Diseases.—The sweet potato is known to be a host for at least one insect and several diseases, which may be introduced into new localities or disseminated where they already occur through the use of "bedding" potatoes that are insect-infested or diseased. The insect is the sweet-potato weevil, and the diseases are black rot, stem rot, scurf, and foot rot. The safest practice is to use for bedding root tubers that are thought to be free from weevils and disease. Then, as an added precaution, the bedding potatoes should be submerged for 10 minutes in a solution of mercuric chloride, prepared by adding 1 ounce to 8 gallons of water.

Potato Diseases.—The Irish potato is subject to a number of serious virous diseases: leaf roll, potato mosaic, witches'-broom, calico, yellow-top, curly top, and others. The symptoms of these diseases vary.

Diseased potatoes when used for planting result in greatly reduced yields and tend to cause the variety to degenerate and "run out." Artificial treatments are not effective in eliminating virous diseases of the potato. A practical method for control is to select tubers for planting from heavy-producing plants that show no evidence of being diseased.

Questions

1. Describe root knot. What is the cause of it? How is it controlled? Name resistant plants.

2. How does grape root louse damage grape plants? What means are used to overcome it?

3. How does crown gall affect plants? What plants are quite susceptible?

4. How does hairy root affect plants?

5. How may plants become affected with crown gall and hairy root?

6. How does collar rot affect plants? What plants are affected by fire blight?

7. What practices are followed in combating it?

8. What is the cause of black end of pear?

9. Name the various virous diseases of the peach. In what ways are they disseminated?

10. What are the principal diseases of the sweet potato that may be disseminated by propagation practices?

11. What diseases of the Irish potato are disseminated in propagation?

Suggested References

1. CUNNINGHAM, H. S.: The Root-knot Nematode in Relation to the Potato Industry on Long Island, *N. Y. (Geneva) Agr. Exp. Sta. Bull.* 667, 1936.
2. DAVIS, L. D., and W. P. TUFT: Black-end of Pears, III, *Proc. Amer. Soc. Hort. Sci.*, **33**: 304–315, 1935.
3. GLOYER, W. O.: Crown Gall and Hairy Root of Apples in Nursery and Orchard, *N. Y. (Geneva) Agr. Exp. Sta. Bull.* 638, 1934.
4. HUTCHINS, LEE M., E. W. BODINE, and H. H. THOMBERRY: Peach Mosaic, Its Identification and Control, *U. S. Dept. Agr., Circ.* 427, 1937.
5. HEALD, F. D.: "Introduction to Plant Pathology," McGraw-Hill Book Company, Inc., 1937.
6. SIEGLER, E. A., and R. B. PIPER: Pathogenesis in the Woolly-Knot Type of Crown Gall, *Jour. Agr. Research,* **43**: 985–1002, 1931.
7. SNYDER, ELMER: Susceptibility of Grape Rootstocks to Root-Knot Nematode, *U. S. Dept. Agr., Circ.* 405, 1936.
8. WATSON, J. R., and C. C. GOFF: Control of Root-Knot in Florida, *Fla. Agr. Exp. Sta. Bull.* 311, 1937.

CHAPTER XVI

TRANSPLANTING

Transplanting consists of moving plants from one place to another with the intention of having them continue their growth in the new location.

The art of transplanting is probably practiced more widely than any other in horticultural work, except that of planting seed. It is important in the growing of flowers, vegetables, and fruits. Many vegetable crops are started in specially prepared seedbeds and later moved to the field. Building sites are quickly made attractive; parks are established; highways are provided with shade; orchard and small-fruit plantations are established; forests are replanted; and flowering plants are rendered more valuable—all by various adaptations of this practice. The distance involved may be small or great, only a few feet or hundreds of miles. Success in either case depends partly upon care exercised in the three rather distinct operations of digging, moving to the new location, and replanting. It depends, also, on the kind of plant, the condition of the plant, and upon certain environmental factors, as, for example, humidity and temperature.

Death or slow growth following transplanting is due mainly to the inability of the plant to take up sufficient moisture through the damaged or disturbed root system. Treatments or conditions that reduce the rate of water loss from the top by transpiration or make it possible for the root system to absorb water more readily increase the chances of survival of the plant.

Methods of Moving Plants.—(1) Most plants are moved *bare-rooted;* the roots are necessarily exposed for a time during the digging, moving, and replanting operations. The exposure and physical injury deprive the roots of practically all root hairs and growing root tips, and some plants do not recover from the shock. (2) Plants may be moved also by *shifting,* an operation whereby plants are started in pots or similar containers, and from these moved, with the soil intact, to a new location, with little or no damage to the root system. This is a means whereby species that do not stand transplanting well are successfully moved. (3) Practically the same results may be obtained for larger plants by *balling and burlapping*. In doing this, the plants are dug so as to include the roots intact in a ball of earth, which is supported by burlap. Balling and burlapping is commonly used in moving evergreen plants. It is also a means whereby deciduous species may be moved during the growing season.

FIG. 165.—Cabbage plants: (1) growing in clay pot, and (2) removed from pot and ready for field planting.

Herbaceous Plants.—Many vegetable and flowering plants are transplanted when in a tender, succulent, growing condition. The success with which such plants can be transplanted depends on several factors.

1. *Formation of New Roots.*—Plants of some species do not stand transplanting well. This is true of corn and many of the peas and beans. It is true also of plants of the cucurbits, such as the watermelon, cantaloupe, and squash. These plants are difficult to transplant because they form new roots slowly and because old roots of cucurbits early develop a suberized layer and are ineffective in the direct absorption of water. Root

hairs are largely lost in transplanting; and, except under most favorable environmental conditions, the plant is likely to die before it becomes reestablished. Such plants can be moved satisfactorily only by shifting. On the contrary, many herbaceous plants can be transplanted readily. This is true of such common vegetables as the tomato, pepper, cabbage, cauliflower, lettuce, onion, and others. It is true also of many flowers as, for example, zinnia, petunia, cornflower, phlox, and verbena.

FIG. 166.—Machine used in transplanting vegetables. Plants are placed into a slowly revolving wheel which releases them in the soil. Three rows are planted at one time. The three double wheels pack the soil around the plants.

These plants are easy to transplant, apparently because they form new roots quickly and are, hence, soon able to supply the top with moisture. This characteristic is especially noticeable in tomatoes; recently transplanted plants will often form new roots by the second or third day following transplanting. In moving a plant from one location to another, it is desirable not only that the plant live, but that it renew growth as quickly as possible.

2. *Hardening.*—Strong, stocky plants that have been properly hardened in the seedbed stand transplanting better than soft,

succulent plants. Hardening is accomplished principally by (*a*) subjecting the plants to relatively lower temperatures and by (*b*) withholding moisture. Neither of these treatments should be carried to the extreme, lest the plant be dwarfed permanently. The object of hardening is to check the growth of the plant to the extent that it may be able to stand adverse conditions after transplanting to the field, such as higher or lower temperatures, wind, dry soil or air, and hot sunshine. In

FIG. 167.—Field of onions established by transplanting from a seedbed.

the process of hardening, the water content of the plants is reduced, and the osmotic concentration increased correspondingly. This condition makes them more retentive of moisture, which is the primary requisite for hardiness to cold, heat, or drought. Hardened plants have a better developed root system in comparison to top growth than is the case with highly succulent plants. Such a root system obviously is able to supply the top more adequately with moisture than is that of a plant not hardened. Furthermore, hardened plants do not lose water by transpiration so rapidly as those not hardened. Hardening is desirable even for plants that are to be shifted from pots to the field.

3. *Care in Handling.*—The care exercised in handling herbaceous plants determines, in a large measure, their response following transplanting. They should be removed from the seedbed with as much of their root system as is practicable and replanted with the least possible delay. They should be protected in the meantime by wet sacks, damp moss, or some other moist insulating material. Often the roots are "puddled," an operation whereby the roots are dipped in a thick mud in order to protect them from excessive drying while they are out of the ground. And, finally, the plants should be replanted in a well-pulverized bed. They should be set slightly deeper than they stood in the seedbed, the soil should be pressed firmly about the roots, and water should be added to settle the soil and increase the amount of available moisture.

4. *Weather Conditions.*—The rate of transpiration is relatively low on cool, moist, cloudy days. The same process normally goes on more slowly late in the afternoon and during the night than during midday. Water requirements are hence less, and the injured root system is able to supply the top more adequately than would be the case if the plant were using more water in transpiration. Thus, plants have a better chance to survive if moved late in the afternoon or on days that are still, cool, cloudy, and humid.

Deciduous Trees and Shrubs.—Excessive transpiration and reduced water supply are the principal causes of death of transplanted plants.

1. *Transpiration.*—Most transpiration goes on through the leaves; therefore a logical time to transplant deciduous plants is during their *dormant* period. It is true that the tree is expected to produce new shoots and leaves when it resumes growth the following spring, but in the meantime it will usually have developed new roots sufficient to supply the entire plant adequately with moisture. Whether such new roots will actually have developed will depend on several factors. It is customary to *cut back the top* of trees and shrubs so as to reduce the amount of foliage produced, and thus restrict transpiration to an amount likely to be supplied by the root system. *Coating the top* of the

tree with melted paraffin, paraffin emulsion, or similar preparations reduces evaporation and the consequent weakening of the top.

2. *Moisture Supply.*—Moisture essential for top growth of plants is absorbed largely by *root hairs* or other very minute

FIG. 168.—Drilling roots of pecan tree to insert toothpicks treated with indolebutyric acid. This treatment has been of value in promoting growth of new roots on transplanted trees. (*Courtesy O. S. Gray, Pecan Nursery, Arlington, Tex.*)

feeder roots. These are ordinarily destroyed when the tree or shrub is removed from the soil. Furthermore, the tips of small and large roots, the regions from which *feeder roots* arise, are destroyed. Thus, new branch roots must arise before new feeder roots can develop. Ordinarily, new branch roots arise from the pericycle of the portion of the root making primary growth, that is, near the growing tip. If the root tips are destroyed, any new roots that form must develop from the cambium of older roots—those that are making secondary growth. These are said to be adventitious roots. Successful transplanting of plants with roots bare depends in a large measure on the formation of such adventitious roots. Some kinds of plants produce such roots readily; others less readily. Differences in

FIG. 169.—Showing response of pecan roots to chemical treatment prior to transplanting. (1) Toothpicks soaked in indolebutyric acid inserted in roots at points where clusters of new roots occur. (2) No treatment.

formation of adventitious roots possibly account for the ease of transplanting the peach and the difficulty encountered in transplanting the pecan. There is some evidence to indicate that adventitious roots normally form more readily on small roots than on larger roots of the same plant. Small roots suffer more from drying and other injury, however. Root pruning during the growing season before a plant is to be moved results in more root branching and a more compact root system. In digging such plants, a greater portion of the root system is obtained than is likely when the practice is not followed.

The formation of adventitious roots may be encouraged or hastened by the use of certain *chemicals*, notably indoleacetic and indolebutyric acid, applied in various ways. Indolebutyric acid has been used successfully in encouraging new root formation in the pecan. Holes are bored transversely into the tap and lateral roots, and toothpicks soaked with 4 milligrams of the acid in solution are inserted in each hole prior to planting. Roots form much more readily at points receiving these treatments than at other places.

New root formation takes place most readily in a *well-aerated soil*. If the soil where the tree is planted is kept waterlogged by rainfall or by irrigation, new root formation is discouraged and the plant is likely to suffer. Plants that have made a normally vigorous growth in the nursery are thought to stand transplanting better than those that have made restricted growth. The reason for this probably is a better supply of *reserve food materials* that encourage a readier formation of adventitious roots.

Two rather distinct practices are followed in the replanting of trees. According to one the tree is placed in the hole slightly deeper than it stood in the nursery. Loose soil is added and pressed firmly about the roots, which are adjusted from time to time in their natural position as far as possible. Sod, clods, and subsoil encountered in digging the hole should be used last in filling in around the tree, and should not be packed in around the roots. It is not advisable to add manure or fertilizer to the recently transplanted tree.

According to another practice, the soil is shoveled in around the roots of the plant, without any effort to pack it. When the hole is almost filled, water is added to settle the soil, after which the rest of the hole is filled. In either case it is important to handle the trees so that the root system is protected against drying or freezing. Puddling the root system with thick mud, as described for herbaceous plants, is a convenient way to protect it against drying.

Evergreen Trees and Shrubs.—Plants that retain their foliage throughout the year are known as *evergreens*. There are two principal kinds. The rhododendron, box, avocado, certain species of ligustrum, and citrus are examples of the so-called "*broad-leaved* evergreen" plants. The pines, cedars, junipers, firs, and arborvitaes are examples of *coniferous* evergreens. In each of these kinds the presence of leaves during the entire year complicates the problem of transplanting. With green leaves on a plant, the rate of transpiration is far greater than it is without them, and the moisture required to keep the plant alive is correspondingly greater. Evergreen plants are not generally

Fig. 170.—Evergreen shrub, balled and burlapped for transplanting. Burlap is wrapped tightly around top to facilitate handling and to protect the branches.

moved bare-rooted, because in most instances the moisture lost by transpiration from the leaves is greater than can be supplied by the injured root system. Death is inevitable if such a condition exists for very long. Two courses of action may be followed: The plant may be *defoliated* in order to lessen the water loss by transpiration and consequently the amount required of the root system. This was formerly an established practice in transplanting citrus, despite the temporary additional dwarfing effect that it had. Although such a practice would be an effective aid in transplanting certain evergreen ornamentals, even temporary defoliation would be objectionable. Such plants are used for their immediate effect in the landscape, and it is desirable that they retain their original appearance. The use of wax emulsions, previously mentioned, has been found of great value in transplanting such material.

The other recourse is to move the plant with a minimum of disturbance to the root system, so that it will continue to function after the plant is moved to its new location. This is done by digging the plants with a portion of the root system undisturbed in a solid ball of earth, which is enclosed tightly in burlap. This operation is practiced widely in commercial work and is known as *balling and burlapping* (B & B). The ball of earth seldom contains all the root system of the plant being moved. Some plants naturally have compact root systems and most of the roots can be included in a ball of reasonable size. Others have long, branching roots that are more difficult to include. Obviously, the size of the plant determines how far out its roots extend from the base. In very poor, sandy soils, roots are likely to be longer and less compact than in a fertile soil. Preliminary root pruning is used effectively in encouraging a compact root system. In all cases, however, effort is made to include enough of the roots in the ball to supply the top with moisture during the period when it is becoming reestablished.

Balled stock may be stored for weeks or months before it is replanted. In the meantime, it should be protected against freezing and particularly drying. Balled stock should always be handled in such a way that the ball of earth is not crushed or broken. When plants are stored in this manner for a considerable period, the sacking deteriorates and the balls often crumble, so that a great amount of reworking becomes necessary.

In replanting it is not necessary or desirable to remove the sack. If the soil is packed closely about the ball the sack will soon decay. The root tips that are not lost or injured continue to grow and give rise to additional feeder roots and to new branch roots that arise in a normal way from the pericycle. The feeder roots in the ball of earth should continue to function in the absorption of moisture as they did before it was dug up. Roots, the tips of which are broken, may give rise only to branch

FIG. 171.—Citrus trees and palms in tree shed for transplanting.

FIG. 172.—Removing a large tree for transplanting.

roots that are of adventitious origin, and these are not formed so freely or quickly by some species as those that arise in normal succession. All these possibilities are closely associated with maintaining for the top a supply of moisture sufficient to prevent permanent wilting, despite a high transpiration rate occasioned by green leaves.

Moving Large Trees.—Very large trees are frequently moved by a process comparable to balling and burlapping. Special transplanting equipment is used and elaborate precautions are

FIG. 173.—Large tree after transplanting. Such trees are pruned moderately and braced securely.

taken to move the plant with an appropriate part of its root system intact and undisturbed in a mass of soil around the base. When this is done it is perfectly feasible to move trees, either deciduous or evergreen, that may be 6 inches or more in diameter.

Treatments that restrict evaporation, such as spraying the top with paraffin emulsion or similar compounds, are considered to be of value in moving large trees. In some cases it is desirable to reduce the top to some extent in order to lessen the amount of water required of the root system.

Questions

1. What is transplanting? Shifting? Balling and burlapping?

2. List the factors that influence success in transplanting herbaceous plants.

3. What is meant by hardening? How is it accomplished?

4. What treatments are used to restrict evaporation from deciduous plants when transplanted?

5. What factors govern the formation of new roots on deciduous plants following transplanting?

6. Why do some deciduous plants resume growth more readily than others after having been transplanted?

7. What are the advantages of balling and burlapping?

8. What other treatments influence the successful transplanting of evergreen plants?

Suggested References

1. BALLANTYNE, A. B.: Fruit Tree Root Systems, *Utah Agr. Exp. Sta. Bull.* 143, 1916.
2. CUMMINGS, M. B., and R. G. DUNNING: A Study in Recovery of Transplanted Apple Trees, *Vt. Agr. Exp. Sta. Bull.* 432, 1938.
3. MULFORD, F. T.: Transplanting Trees and Shrubs, *U. S. Dept. Agr., Farmer's Bull.* 1591, 1929.
4. TALBERT, T. J.: Transplanting Fruit Trees, *Mo. Agr. Exp. Sta. Bull.* 245, 1927.
5. WHITTEN, I. C.: An Investigation in Transplanting, *Mo. Agr. Exp. Sta. Research Bull.* 33, 1919.
6. WOODROOF, J. G.: Relation of the Root System of Pecan Trees to Nursery and Orchard Practice, *Ga. Agr. Exp. Sta. Bull.* 176, 1933.
7. ——— and NAOMI C. WOODROOF: Pecan Root Growth and Development, *Jour. Agr. Research* **49**: 511–530, 1934.

CHAPTER XVII

GROWING AND HANDLING NURSERY STOCK

The routine followed in the commercial production and handling of nursery stock suitable for permanent planting varies with different species and also with the geographic locality. The methods used in producing the stock may influence the later growth of a plant in many ways. For this reason it seems appropriate to consider certain factors of nursery practice and management.

Nursery Site.—There are many factors involved in the choice of a location for a nursery. The most important of these have to do with the soil, slope of the land, fertility, rainfall, and temperature conditions; facilities for transportation must also be given consideration.

1. *Soil.*—The different parts of the nursery enterprise will vary somewhat with regard to soil requirements. Seedbeds are best located on rather light, sandy loams, which do not bake or form a hard crust and are easy to cultivate. Small stock that is to be transplanted soon should also be grown on a light soil. On the other hand, trees that are to be grown several years will need a stronger soil; this is especially true of evergreens, some of which remain in the nursery for a number of years. A moderately heavy soil is considered by many nurserymen to be more desirable for balling.

The most important characteristic of a good nursery soil is structure. It should extend to a depth of 18 to 24 inches without noticeable hard layers. Only in a soil of this type will nursery stock produce a satisfactory root system. In the Kaw River valley of Kansas, where many of the apple stocks are grown, the soil is alluvial and is uniform in structure to a great depth. The lower levels of the soil or subsoil should also provide adequate drainage.

2. *Slope of the Land.*—It is generally considered that nursery land should have little or no slope. Level land is frequently

more uniform, especially where it has been deposited in a river bottom. Such land is easier to irrigate, by either the furrow method or an overhead sprinkler system. It is also easier to cultivate, since it will not be subject to erosion.

3. *Fertility.*—Soil for the growth of nursery stock should be at least moderately fertile. A poor soil has a tendency to cause a widespread root system with comparatively few branch roots. Such a root system is unsatisfactory for transplanting; a close, compact set of roots is considered most desirable. Trees and

FIG. 174.—Large block of perennial Phlox, growing on level land, and with facilities for sprinkler irrigation. (*Courtesy Shenandoah Nursery, Shenandoah, Iowa.*)

plants that are to remain on the land only one season will not be affected so much as those whose growth will extend over a period of years.

The physical condition of the land, especially from the standpoint of texture and organic matter, should be extremely good at the time it is planted to nursery stock. It is impossible to grow cover crops in the nursery, but a good cover crop should precede its planting. Manure and other organic materials may be plowed under and allowed to decompose, to aid in the preparation of the land.

4. *Climate.*—The principal climatic factors involved in the growing of nursery stock are rainfall, temperature, and length of growing season. A moderate amount of rainfall, distributed throughout the growing season, will cause a more uniform growth of the plants. High summer temperatures, especially when accompanied by lack of rainfall, will make irrigation desirable. Winter temperatures may also be of considerable importance. Early freezes, before the deciduous trees are fully dormant, may cause damage. On the other hand, continued cool weather in

FIG. 175.—Large block of 3-year peonies in nursery. (*Courtesy Shenandoah Nursery, Shenandoah, Iowa.*)

the fall will aid in maturing and hardening the trees, so that they may be dug and placed in storage.

Cultural Practices.—Nursery stock is planted closely and makes full use of the land within a short time. The fertility of the soil is considerably depleted by a crop of nursery trees. If they remain on the land for several years it will be necessary to make applications of commercial fertilizer to maintain proper growth. Such applications, especially of nitrogenous fertilizers, should not be made late in the summer, as they tend to prolong growth and delay maturity of trees.

Clean culture should be practiced to keep down weeds and grass and maintain the moisture supply. Frequent shallow cultivation will give best results. A considerable amount of hoe work will also be necessary to keep down the growth of vegetation in the tree row. Spraying for the control of insects and diseases is usually necessary. Most nursery stock is sold on the basis of size, either trunk caliper or height; proper growing

FIG. 176.—Spray rig used for nursery trees.

FIG. 177.—Catalpa Bungei in nursery row, showing well trained trunks. (*Courtesy Shenandoah Nursery, Shenandoah, Iowa.*)

conditions throughout the entire season will produce more plants in the higher grades.

Pruning and Training.—The type of pruning in the nursery will vary with different plants. Fruit-tree stocks in which a single, straight stem is desired will require the removal of sprouts and extremely low branches. One-year "whips" will usually grow straight, but two-year branched trees will frequently have more branches than are desired. Citrus trees are commonly headed at the end of one season, and the top trained to a few scaffold branches during the second season. Ornamental plants of many kinds are headed and pruned to give them desirable shapes. In extreme cases they are sheared so as to become quite compact, and they are also cut into various formal shapes.

FIG. 178.—Lombardy poplar in nursery. (*Courtesy Shenandoah Nursery, Shenandoah, Iowa.*)

Root pruning is practiced in many cases in order to cause the formation of a dense, compact root system, better adapted to transplanting. It is usually practiced upon young plants, only 1 or 2 years in the nursery, to prevent too great a spread of lateral roots, or to restrict the growth of the taproot. On the other hand, root pruning is frequently practiced on older plants 1 or 2 years before they are transplanted. In this case a trench is dug around the plant as if it were to be balled and transplanted; but the trench is then refilled. The severed roots have a tendency to make a very compact growth near the point where they were cut. When the plant is dug, the ball is made sufficiently large to include this new branched root system.

Digging.—The majority of all nursery stock is dug in the fall, when the plants have completed their growth and the new wood has become mature. A considerable number of trees and other stock are sold at this time for fall planting; on the other hand, much of the stock is placed in storage, where it will be available for delivery in the spring. In either case it is desirable that the plants be in a mature condition for digging; immature wood will not be satisfactory for fall-planted trees or for those to be stored.

Some kinds of deciduous nursery stock may be dug with a specially designed, U-shaped blade, to which are attached handles for guiding. The machine is drawn down a tree row by

FIG. 179.—A nursery-tree digger.

horses or by tractor. The blade severs the taproot at the desired depth, 12 to 16 inches, and the lateral roots, except those in the tree row. The trees may then be lifted out of the ground and the soil shaken from the roots.

A common practice is to dig bare-rooted plants with special nursery spades, which are built with strong reinforcement. Two men, working on opposite sides of the tree row, can sever the long roots around a tree and lift it out with the greater portion of the root system intact.

Frequently, following a season of favorable growth, the trees will be in full leaf at the time it is advisable to dig them. This is especially true of one-year seedling apples that are to be used for whip grafting or lining out in the nursery for budding. Such trees are lifted in full foliage as soon as cut by the tree digger, tied in bunches, and placed in deep furrows, which are covered at once. In this green, moist condition the trees will sweat and shed their leaves in a short time. After a period of 2 to 3 weeks, they are removed and carried to the shed or cellar, where they are to be graded.

The various ornamental evergreens and trees of citrus and subtropical fruits are usually dug with a ball of earth, which is

FIG. 180.—Variation in rose plants grown in same nursery row, showing the need for grading standards. (*Courtesy A. F. Watkins, Dixie Rose Nursery, Tyler, Tex.*)

wrapped in burlap. Plants of this kind are allowed to remain in the nursery until spring and are usually dug as demand for them requires. Balled nursery stock is heavy to handle and not entirely satisfactory to store for long periods. In regions where winter temperatures are not often dangerous, citrus trees and broad-leaved evergreens are stored for short periods in slatted sheds. They may be protected against sudden cold by the use of canvas flaps on the sides and top of the shed or by the use of heaters.

Tree Grades.—As soon as the trees are dug, if they have shed their leaves, they are carried into a workroom or shed of some kind where they are sorted according to grade. There are specific grades for all kinds and varieties of fruit trees and other nursery stock. These standards are set up by the American Association of Nurserymen. Two-year apple and pear trees are graded according to caliper, with height a secondary consideration. Common sizes are $^{11}/_{16}$ to 1 inch, $^{9}/_{16}$ to $^{11}/_{16}$ inch, $^{7}/_{16}$ to $^{9}/_{16}$ inch, and $^{5}/_{16}$ to $^{7}/_{16}$ inch. They may be designated as large, medium, and small; or XXX, XX, and X. Roses, which are frequently quoted as two-year, field-grown plants, are also graded as Nos. 1, 1½, 2, and 3, with a "Supreme"

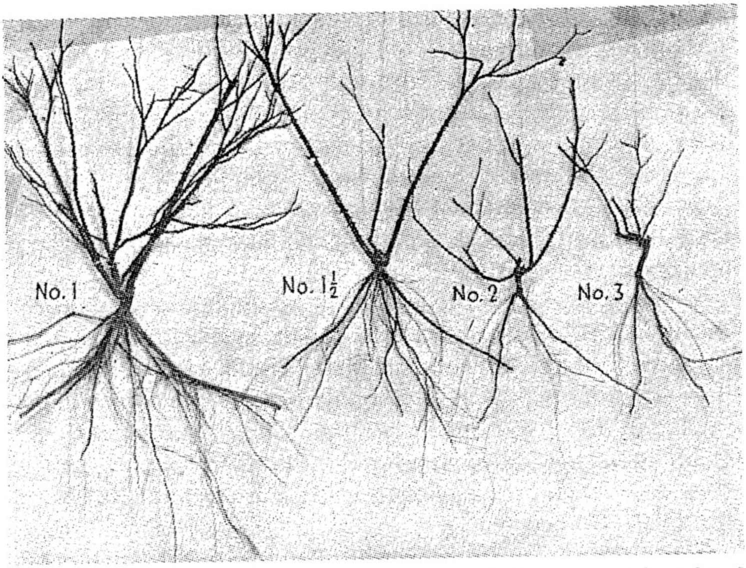

FIG. 181.—Single plants showing characteristics of the commercial grades of roses. (*Courtesy A. F. Watkins, Dixie Rose Nursery, Tyler, Tex.*)

grade sometimes placed above No. 1. Size of plants and number of strong canes are important factors in rose grades.

There has always been considerable difference of opinion as to size and age of trees that are to be planted in the orchard. Most of this discussion is concerned with the relative merits of one- and two-year trees, or of June buds in the case of peaches. It is reasonable to assume that, within a given lot of nursery trees, the individuals that have made the best growth will also give the best response in the orchard. Trees of the lower grades are purchased by some growers on account of the difference in price. The cost of the trees, however, is a rather small item of the total cost of bringing an orchard up to bearing age. Saving in the purchase price of trees may not always be true economy.

Seedling fruit-tree stocks, as of apple and pear, are graded into ¼, $^{3}\!/_{16}$ (No. 1), and $^{2}\!/_{16}$ inch (No. 2); each of these grades is also divided into plants with branched roots and those with straight taproots. The branched roots, used for whole-root grafting or for lining out as budding stock, sell at a higher price than the straight roots, which may be used for piece-root grafts. Seedling cherry, peach, plum, and rose stocks, as well as rooted cuttings of quince, are graded in similar manner by caliper.

Storage.—Conditions in the fall are usually favorable for digging of nursery stock. The soil does not get cold until late in the year and is not likely to be too wet for the removal of the trees. In the springtime, on the contrary, the soil thaws slowly and is also too wet for nursery work. For this reason, the fall is the season when most of the nursery stock is dug. A certain percentage of the stock is sold in the fall, but much of it is placed in storage.

A storage for nursery stock must be of ample size and must be constructed so as to protect the plants against extreme temperatures. In the fall when the plants are first stored, temperatures are often higher than desirable to keep the trees in good condition. On the other hand the winter temperatures are often low enough to cause damage to the stored trees, especially the roots. Careful construction and good insulation are essential if the storage house is to be of greatest value. Proper provision must be made for placing the trees and other stock in bins and racks, where the different varieties and grades will be accessible. There should be ample space in the workrooms for grading and bundling the stock for storage and also for packing stock for shipment.

The bundles of trees are packed or "corded" very compactly in the bins, and the roots are covered with damp shingle tow or other packing material. It should be kept in mind that a tree is not likely to be so good after a long storage period as when first dug; but long experience has shown that trees that are properly cared for may be stored successfully.

Wax Emulsions for Storage of Plants.—Wax emulsions have been found very useful in handling, storage, and transplanting of nursery stock. In Michigan a "standard emulsion" that contains approximately 13.8 per cent solid material was found to be definitely beneficial to dormant fruit trees. The trees and suitable checks were later transplanted, and those that were sprayed with the wax emulsion while the trees were heeled-in began growth faster than the unsprayed trees and recovered more rapidly after a late freeze that killed back new growth. The emulsions were found to be of great value in retarding needle fall from Norway spruce trees that were cut and handled as for Christmas trees.

Fumigation.—Many serious insect pests have been introduced into this country, and later spread over large areas, on nursery stock. At the present time the different states have laws with regard to inspection of plant material in the nursery; permits are issued when the stock is found to be free of injurious pests. Unfortunately, some insects, as well as many diseases, are rather difficult to detect in certain stages. For this reason, the Federal government has established quarantines to prevent the movement of plant material from certain areas of known infestation. A very stringent quarantine is also in force with regard to importations from foreign countries. In many cases, however, fumigation of the stock is recognized as an acceptable insurance against certain pests. Many states require fumigation with cyanide gas of stock that may carry San Jose scale.

The fumigation chamber may be of any convenient size, depending on the quantity of material to be treated. In some cases, as in the railway points of entry from Mexico, houses are provided that will permit fumigation of a freight car and contents. Such houses are used for plant products of all kinds, rather than for nursery stock. The walls of the fumigation chamber should be constructed of tongue-and-groove lumber, or other material that will retain the gas.

Nursery stock for fumigation should be dormant and should not be wet. Open bundles are piled in the chamber loosely or stacked crosswise so as to permit movement of the gas. The dosage in the tabulation is sufficient for 100 cubic feet of space. The time of treatment is 40 minutes.

Sodium cyanide	1 ounce
Commercial sulphuric acid (sp. gr. 1.84)	1½ fluid ounce
Water	2 fluid ounce

Water is placed in an earthenware jar or crock, acid is added to the water, and the cyanide, in paper bags, is added last. If several jars are used they are all prepared in advance, and the cyanide placed in rapidly by the operator, who closes the chamber immediately. Cyanide gas is dangerous, and every precaution should be observed in its use. It is never advisable for one man to work alone in such an operation.

Packing for Shipment.—Specimen plants, or other balled-and-burlapped trees, cannot be placed in compact packages. Balled trees are sometimes placed in crates, with braces to hold

FIG. 182.—Packing room of a nursery. (*Courtesy A. F. Watkins, Dixie Rose Nursery, Tyler, Tex.*)

the individual specimens in place. In the case of large shipments, however, such plants are packed closely into the car or truck, without the use of crates, and are braced securely in place.

Bare-rooted plants are packed into bundles for shipment, and special provision is made to protect the roots from drying out. The plants are first tied in bundles, which are drawn together compactly by means of a machine made up of several sets of semicircular clamps. The roots are frequently "puddled" by dipping them in a thick mud in order to protect them against excessive drying. The roots are packed in damp shingle tow or sphagnum moss for shipping. The tops of the plants may be protected by means of rushes or similar material, which is also tied in place. The bundle is well wrapped in heavy paper, specially designed to prevent loss of moisture. The whole package is then wrapped tightly with burlap, which is sewed with jute, binder twine, or similar material. Such a package will protect the plants from drying for a considerable time and will also protect the tops. Packages for mail must be kept within requirements, but large bales are commonly packed for freight or express shipment. Large shipments of bare-rooted plants are frequently packed in wooden cases, which are lined with waterproof paper.

Questions

1. What factors influence the choice of a location for a nursery?
2. What influence does the fertility of soil have on the root system of a nursery tree?
3. What determines the type of pruning and training that nursery plants receive?
4. What is the reason for the practice of root pruning nursery plants?
5. What special equipment is used in digging plants?
6. How is nursery stock stored?
7. What is the purpose of fumigation? How is it accomplished?
8. What are the steps in packing nursery stock for shipment?

References

1. KAINS, M. G., and L. M. MCQUESTEN: Propagation of Plants, Orange Judd Company, 1938.
2. LAURIE, ALEX, and L. C. CHADWICK: The Modern Nursery, The Macmillan Company, 1931.
3. TUKEY, H. B., and CARL D. BRASE: The Response of Apples, Cherries and Roses to Fertilizer Applications in the Nursery, *N. Y. (Geneva) Agr. Exp. Sta. Bull.* 509, 1931.

INDEX

Page numbers in **bold-faced** type indicate main references.